SpringerBriefs in Earth Sciences

SpringerBriefs in Earth Sciences present concise summaries of cutting-edge research and practical applications in all research areas across earth sciences. It publishes peer-reviewed monographs under the editorial supervision of an international advisory board with the aim to publish 8 to 12 weeks after acceptance. Featuring compact volumes of 50 to 125 pages (approx. 20,000–70,000 words), the series covers a range of content from professional to academic such as:

- timely reports of state-of-the art analytical techniques
- bridges between new research results
- snapshots of hot and/or emerging topics
- literature reviews
- in-depth case studies

Briefs will be published as part of Springer's eBook collection, with millions of users worldwide. In addition, Briefs will be available for individual print and electronic purchase. Briefs are characterized by fast, global electronic dissemination, standard publishing contracts, easy-to-use manuscript preparation and formatting guidelines, and expedited production schedules.

Both solicited and unsolicited manuscripts are considered for publication in this series.

More information about this series at http://www.springer.com/series/8897

Khaini-Kamal Kassymkanova ·
Gulnara Jangulova · Gulnura Issanova ·
Venera Turekhanova · Yermek Zhalgasbekov

Geomechanical Processes and Their Assessment in the Rock Massifs in Central Kazakhstan

Khaini-Kamal Kassymkanova
Faculty of Geography and Environment
Al-Farabi Kazakh National University
Almaty, Kazakhstan

Gulnura Issanova
Faculty of Geography and Environment
Research Center for Ecology
and Environment of Central Asia
Al-Farabi Kazakh National University
Almaty, Kazakhstan

Yermek Zhalgasbekov
Faculty of Geography and Environment
Al-Farabi Kazakh National University
Almaty, Kazakhstan

Gulnara Jangulova
Faculty of Geography and Environment
Al-Farabi Kazakh National University
Almaty, Kazakhstan

Venera Turekhanova
Faculty of Geography and Environment
Al-Farabi Kazakh National University
Almaty, Kazakhstan

ISSN 2191-5369   ISSN 2191-5377   (electronic)
SpringerBriefs in Earth Sciences
ISBN 978-3-030-33992-0   ISBN 978-3-030-33993-7   (eBook)
https://doi.org/10.1007/978-3-030-33993-7

This Springer imprint is published by the registered company Springer Nature Switzerland AG
The registered company address is: Gewerbestrasse 11, 6330 Cham, Switzerland

Preface

The Republic of Kazakhstan is one of the largest republics of Central Asia, with an area of more than 2.7 million square kilometers, a population of more than 18 million people, land borders with Russia, China, Uzbekistan, and Kyrgyzstan. The bowels of Kazakhstan store a significant number of types of minerals. On the basis of explored reserves, a powerful mining industry has been created, for the extraction and processing of ores of ferrous, nonferrous, and noble metals of various types. The mining and metallurgical complex includes more than 200 mining and processing enterprises, the sale of marketable products of which is currently carried out in European countries, the USA, China, South Korea, Singapore, Malaysia, and other countries.

Mining is the most important branch of material production, the level of which determines the economic power and independence of states. Mining provides countries with the necessary natural resources, on the basis of which the manufacturing industry creates means of labor and consumer products.

Mining projects provide the population with work, build various infrastructure objects, and raise the living standards of the population.

In the field of open-pit mining of mineral deposits, the scientific potential of this area is of particular importance in the design and long-term planning of mining in deep pits, which currently produce more than 90% of iron ores, at least 35% of nonferrous metals, about 8% of coal in Kazakhstan , about 70% of asbestos and phosphate ores.

A major perspective area of research in the field of open-pit mining is the management of the developing of mining operations. This direction includes knowledge of methods for managing the development of mining operations, design methods, and substantiation of the regularity of mining development in time and space, research of geomechanical and geotechnical processes.

The aim of the research is to increase the stability of the quarry boards, the slope of the ledges by strengthening and hardening them for the effective and safe development of mineral deposits. New methods for determining the disturbance of the mountain massif are studied in detail and proposed in this work.

The significance of the results is the implementation of a complex of investigations of a field tectonics crack developed by the open method, which takes into account thermal, ultrasonic methods for strengthening the sideboards of the quarry, formation, and processing point diagrams from field measurement data and a map of the fracture of the most unfavorably oriented slopes. To improve the efficiency of its processing, the development of methods for express assessment of disturbance of the mountain massif for its operational control, strengthening of quarry slopes composed of fractured rocks has been proposed.

The monograph is of interest to mining engineers, researchers, undergraduates, and students of geomechanical, geodesic, mining, and environmental specialties.

Almaty, Kazakhstan   Khaini-Kamal Kassymkanova
                                                                                  Gulnara Jangulova
                                                                                  Gulnura Issanova
                                                                             Venera Turekhanova
                                                                             Yermek Zhalgasbekov

Acknowledgements

Many scientists have contributed, supported, and provided the basis for this book. We sincerely thank all of the scientists: Dr. Nurpeisova Marzhan, Dr. Mashanova Alisher (Head of the Department of Geomechanical Processes, the Kunaev Institute of Mining). Tulebaev Kairat and Omiralin Murat (researchers at the Kunaev Institute of Mining), Bektur Bakytbek (Ph.D. student of the Satpaev University), who had provided assistance and advice during the writing of this book, sharing their knowledge, publications of which became part of the chapters.

Special thanks to Dr. Bekmurzaev Batyrkhan for providing material on geoinformation technology in the design and planning of open-pit mining.

The authors sincerely thank all the participants and contributors in the project, colleagues, friends, and well-wishers for their support.

This research was funded by the Ministry of Education and Science of the Republic of Kazakhstan for 2015–2017, (grant 0424/GF 4) "Express-assessment of geomechanical condition of the rock massive and development methods of its strengthening and reinforcing for safe ecological developing of the fields of mineral resources in hard mountain—geological and mining engineering conditions."

Content and Structure of the Book

This monograph is the result of research on grant financing "0424/GF 4 Express-assessment of geomechanical condition of the rock massive and development methods of its strengthening and reinforcing for safe ecological developing of the fields of mineral resources in hard mountain—geological and mining engineering conditions". Issues for the environmentally safe mining of mineral deposits were reviewed to develop methods for: rapid assessment of disturbance of the mountain massif; fortifications; hardening quarry slopes; dust suppression on quarry roads.

Chapter 1, "Mineral Deposit and Studying of World Experience on the Study of the Geomechanical State of the Mountain Massif in Complex Mining and Geological Conditions." The issues of the development of mineral deposits, developed by the open method, characterized by a large variety of geological, mining, geomechanical, and technological conditions. Leading enterprises of the mining complex of Kazakhstan. Chapter 2 "Geological Setting Structural-Tectonic Features and Physico-Mechanical Properties of Mountain Rocks on the Stability of Slopes with the Account of the Time Factor and Large-Scale Explosions." This chapter discusses factors affecting on the stability of slopes with including rocky and semi-rock formations that must be taken into account when studying geomechanical processes. Chapter 3 "Petrology and Geochronology of Country and Studying of the Disturbance of Rock Massif with the use of the Teplometry Method." The main rocks and minerals at the Konyrat deposit are of three types: granodiorites, porphyrites, secondary quartzites. Chapter 4 "Spatial Distribution of ore Bodies and Teplometric Method of Express Evaluation of the Rock Massif Disturbance". A heat-measuring method was chosen to carry out an express assessment of disturbance in mountain ranges, which is distinguished by accuracy, measurement speed with ease of work, determination of the boundary of homogeneous zones according to the degree of disturbance of the pit slope, and on the basis of these data it is possible to correct drilling and blasting operations on the section adjacent to the pit slope supplied design position that will increase the stability of the pit walls for the efficient and safe mining of mineral resources in the bottom harboring horizons. Chapter 5 "Ultrasonic Method of Express Evaluation of the Rock Massif Disturbance." The ultrasonic method of research is used, which has a high sensitivity to structural changes in the massif at the early stages of deformation development. Based on the determination of the propagation velocity of longitudinal and transverse elastic waves, samples of using appropriate ultrasonic equipment and measurement techniques were carried out. Chapter 6 "Development of Solutions for: Hardening, Strengthening the Sides of the Quarry; Suppression Dust Formation on the Quarry Roads." The composition of cementing solutions for hardening, strengthening of quarry slopes and dust suppression on the roads of the quarry has been developed. It is based on the task of creating a compound for fixing the dusting surfaces of the open-pit roads using waste from the mining and processing industry, the coating of which has high strength and weather resistance. Chapter 7 "Modern Methods of Researching of the Geotechnical State of the Massif for Engineering and Planning of Open Mining Works". As a result of analyzing various options for the structure and functions of the system and its elements, a model of computer-aided design and planning was built, which most closely corresponds to the conditions for implementing an automated decision-making technology for openwork tasks.

Introduction

The mineral deposits that developed with open mining works are characterized by a wide variety of geological, mining, geomechanical, and technological conditions. The leading enterprises of the mining complex of Kazakhstan are working out the fields by the open method. About 35% of copper ore is mined at the "Nikolayevsky" and "Shemonaikhinsky" open-pit quarries of the East Kazakhstan copper combine, the "Tur" mine ("Kazchrome") and the "Konyrat" mine of the Balkhash Mining and Metallurgical combine.

An important stage in the development of open-pit mining of mineral deposits is characterized by the following features: increasing the depth of the quarries, the lifetime of the slopes of the ledges and boards of the quarry, the growth of over-burden, intensification and concentration of mining, the complexity of engineering, geological and hydrogeological conditions of mining, low content of useful components in the ore. Over 70% of quarries have a depth of over 200 m, many quarries work out the horizons of 400–500 m from the earth's surface, and the design depths reach 700 m or more. In order to improve the efficiency and completeness of field development, improve the technical and economic indicators of the enterprise, to ensure the safety of mining operations in the open pit, reliable maintenance of the stability of quarry slopes is required. In this case, the main task is to determine the optimal parameters of the slopes, ensuring their long-term stability with the minimum amount of overburden works. In addition, the control and regulation of the behavior of the array of mineral deposits in the development require a deep study of the properties and structure of the array, its geological, geostructural, and geotechnical characteristics, as well as geomechanical features.

The aim of the research is to increase the stability of the quarry boards, the slope of the ledges by strengthening and hardening them for the effective, and safe development of mineral deposits. New methods for determining the disturbance of the mountain massif are studied in detail and proposed in this work.

The significance of the results is the implementation of a complex of investigations of a field tectonics crack developed by the open method, which takes into account thermal, ultrasonic methods for strengthening the sideboards of the quarry, formation and processing point diagrams from field measurement data, and a map of the fracture of the most unfavorably oriented slopes. To improve the efficiency of its processing, the development of methods for express assessment of disturbance of the mountain massif for its operational control, strengthening of quarry slopes composed of fractured rocks has been proposed.

The monograph is of interest to mining engineers, researchers, undergraduates, and students of geomechanical, geodesic, mining, and environmental specialties.

Considering the importance of stakes, the slope design in open-pit mines must be based on a well-controlled methodology, especially since experience shows that each rock mass characterized by its geological structures is unique, and therefore there are no standard recipes that achieve the right solution with certitude. This methodology can be broken down into several phases: (1) characterization of the rock mass through the acquisition and analysis of geological and geomechanical data, (2) identification n of potential mechanisms of deformation and failure, and their modelling; (3) the slope design and the definition of methods of reinforcement and monitoring. These phases largely developed by Cojean and Fleurisson are briefly recalled here.

This phase involves the acquisition of geological, geomechanical, and hydrogeological knowledge by observation and measurement. It employs all the disciplines of earth sciences and mechanical sciences, and particularly the disciplines of the engineering geology, geotechnics, soil and rock mechanics, and hydrogeology and groundwater hydraulics. First, the geological approach is essential in order to analyze material behavior. The geologist identifies the petrographic nature of the material (rock or soil) and their state of weathering and fracturing. These data are essential for the characterization of mechanical properties of material. It also provides the spatial variability of these parameters throughout the mass. Similarly, the geologist identifies geological structures of the deposit.

The data obtained from this initial geological approach are significant because they will then guide and optimize the geological and geotechnical field investigations using subsurface geophysical methods, drilling operations or shallow excavations carried out with hydraulic excavator which can, cost-effectively, provide valuable information.

Particular attention must be given to the discontinuity network that cuts the rock mass at different scales. Natural variability of the geometric but also mechanical parameters of the discontinuities requires statistical study and therefore the implementation of rigorous sampling methods. They include the following stages field measurements of discontinuities through systematic survey on outcrops, excavation face or oriented core drilling; classification of discontinuities in directional sets using stereographic projection techniques or automatic classification; statistical analysis of the geometrical parameters of each set using histograms of the principal geometric characteristics of the discontinuities: dip direction, dip angle, persistence, or trace length and spacing.

Contents

About the Authors

Khaini-Kamal Kassymkanova holds a Doctorate degree of Technical Sciences, Associate Professor, Head of the Department of Cartography and Geoinformatics of Al-Farabi Kazakh National University (Almaty), Kazakhstan.

She graduated from the Kazakh Polytechnic University in Kazakhstan (Almaty). She had been working at the enterprises of Kazakhmys Corporation LLP, Balkhash, Kazakhstan in the position of surveyor engineer. Then she continued her studies at the doctoral degree at the Kazakh Polytechnic University, and after graduation she obtained a degree of Doctor of Technical Sciences and is associate professor in the field of mine surveying and geodesy.

Her research interests are focused on the study of the stability of mountain ranges in the development of mineral deposits in difficult geological conditions by the open method at great depths.

She regularly participates in international conferences, symposia with reports on the studies conducted and the results obtained on the sustainability of the career slopes of solid mineral deposits developed by the open method in Central Kazakhstan. The scientific results obtained are patented in the Republic of Kazakhstan.

Khaini-Kamal Kasymkanova has a sufficient number of articles in international journals that are part of the Scopus database, the Russian Science Citation Index (RISC) and is the author and co-author of several monographs and textbooks, such as Innovative Geomonitoring Techniques for Predicting Hazardous Technogenic Phenomena during Subsoil Development,

"Designing re-underground technology in the development of natural and man-made reserves of the collapsed deposits of the Zhezkazgan field", as well as the author of one of the chapters in the monograph ecological and Industrial Safety Exploitation of Mineral Resources published in Russian in Kazakhstan.

For achievements in scientific and educational activities, she was awarded the title "The best teacher of the university of Kazakhstan Republic."

Gulnara Jangulova holds a Candidate degree of Technical Sciences, Associate Professor of the Department of Cartography and Geoinformatics of Al-Farabi Kazakh National University (Almaty), Kazakhstan.

She graduated from the Kazakh Polytechnic University in Kazakhstan (Almaty). She had been working at the Irtysh Polymetallic Combine (Kazakhstan), as a mining engineer—surveyor. Conducted industrial tests and participated in the implementation of developments in enterprises: JSC "Kazchrome", LLP "Kazakhmys".

She had been working at the Kunaev Institute of Mining as a leading engineer of the laboratory, and upon graduation she got a candidate degree in engineering sciences and an Associate Professor in the field of mine surveying and geodesy.

Her research interests are focused on the study of the stability of mountain ranges in the development of mineral deposits in difficult geological conditions by the open method at great depths.

She regularly participates in international conferences, symposia with reports on the studies conducted and the results obtained on the sustainability of the career slopes of solid mineral deposits developed by the open method in Central Kazakhstan. The scientific results obtained are patented in the Republic of Kazakhstan.

Jangulova Gulnara has a sufficient number of articles in international journals included in the Scopus database, the Russian Science Citation Index (RISC) and is the author and co-author of several monographs, textbooks, such as "Innovative geomonitoring methods for predicting dangerous technological phenomena when developing the subsoil," "Designing re-underground technology in the development of natural and man-made reserves of collapsed deposits of the Zhezkazgan field," as well as the author of one of their chapters in the monograph.

Advanced training: refresher course on accreditation and rating, familiarization course on the use of modern geodetic instruments, a course on the quality of education in higher education within the framework of the International Symposium "Quality of education and accreditation in higher education: challenges of the 21st century," and EU ERASMUS courses.

Gulnura Issanova holds a doctorate degree in Natural Sciences and is an Associate Professor at the Al-Farabi Kazakh National University, scientist and researcher at U.U. Uspanov Kazakh Research Institute of Soil Science and Agrochemistry and a scientific secretary at the Research Centre of Ecology and Environment of Central Asia, Almaty, Kazakhstan.

She studied at the Al-Farabi Kazakh National University for bachelor's degree (B.Sc.) and master's (M.Sc.) degree in Physical Geography and Xinjiang Institute of Ecology and Geography, Chinese Academy of Sciences, China, for her doctoral degree. She did a postdoc under the CAS President's (Bai Chunli) International Fellowship Initiative (PIFI) for 2017–2018 at the Xinjiang Institute of Ecology and Geography, Chinese Academy of Sciences, China. She is a holder of the Foreign Expert Certificate of the People's Republic of China. Currently, she is a postdoc at the Al-Farabi Kazakh National University, Faculty of Geography and Environment.

Her research interest was focused on problems of soil degradation and desertification, in particular, the role of dust and sand storms in the processes of land and soil degradation and desertification during the Ph.D. study. Currently, she is interested and focusing on water resources and lakes (availability, state, and consumption/ use) and Aeolian processes in Central Asian countries.

She participates regularly in the International Scientific Activities (Conference, Forum, and Symposium) on Environmental Problems as well as writes articles and monographs on the subject and takes part in local and international projects. She has published many SCI papers in international peer-reviewed journals with high level and wrote a handbook, "How to Write Scientific Papers for International Peer-Reviewed Journals" (in Russian and Kazakh languages).

She is an author of the monograph *Aeolian processes as dust storms in the deserts of Central Asia and Kazakhstan* by Springer Nature (2017) and co-author of *Man-Made Ecology of East Kazakhstan* by Springer Nature (2018), *Hydrology and Limnology of Central Asia* by Springer Nature (2019), and *Overview of Central Asian Environments* (in Chinese) and the handbook *Methodical Handbook on Interpretation of Saline Soils* (in four languages: Kazakh, Russian, English, and Chinese).

She became a laureate of the International Award "Springer Top Author" and was awarded in the Nomination "Springer Young Scientist Awards-2016" for high publication activity in scientific journals published by Springer Nature. She was included to the list of "Top-8 Young Scientists and Top-18 leading Scientists from Kazakhstan," 2016.

Venera Turekhanova is a Master of Natural Sciences, and a second-year doctoral student (Ph.D.) at the Department of Cartography and Geoinformatics of the Al-Farabi Kazakh National University. She holds a bachelor degree in the field of studying physical processes, their influence on various constructions, their application, and received qualified knowledge in the field of mathematical modeling at the Faculty of Mathematics and Mathematics, Al-Farabi Kazakh National University (Kazakhstan, Almaty).

She had an internship at Berlin Technical University, Berlin (Germany), where she attended a monthly course of lectures by professors in mathematics, physics and mechanical engineering, as well as technical mathematics.

Her field of interest focuses on the study of physical laws and the possibility of their useful application, working with modern satellite data, the possibility of predicting various indicators from these data, and the possibility of their application in science.

She worked at the Scientific Research Institute of Mechanics and Mathematics as a lab technician, where she carried out researcher work, as well as engaged in science in the field of mathematics and computation. For several years, she had been working as a laboratory assistant at the Kunaev Institute of Mining.

She has published many articles, and a patent was issued in co-authorship, as well as participated in international conferences. She had an internship-training course on programming by LaVACCa Autumn School organized by the University of Würzburg (Germany) and Urgench State University (Uzbekistan).

Yermek Zhalgasbekov is a Master in technics and technology, and a second-year doctoral student (Ph.D.) at the Department of Cartography and Geoinformatics of the Al-Farabi Kazakh National University.

He holds a bachelor degree in mining, a study on the basis of full-scale measurements and subsequent geometric constructions of the structure of the field, as well as the reflection of the dynamics of the production process of the mining enterprise at the mining and metallurgical Satpayev Institute of Kazakh National Technical University (Kazakhstan, Almaty).

His field of interest is focused on the study of mining planning, using modern equipment and satellite data, the possibility of forecasting various indicators on these data, as well as the possibility of their application in science.

Yermek Zhalgasbekov worked as an engineer at the Kunaev Institute of mining, where he performed research work, as well as engaged in science in the creation of three-dimensional geological model of mineral field. He has extensive experience with modern surveying equipment and programs for processing the results of measurements. He has published about 20 articles and patents. Participated in national projects and international conferences.

Chapter 1
Mineral Deposit and Studying of World Experience on the Study of the Geomechanical State of the Mountain Massif in Complex Mining and Geological Conditions

Abstract At present, the specific weight of the open method of mining mineral deposits (75% of the total world mining) attests to the preservation of this general direction for the development of mining industries to provide fuel and mineral raw materials.

At present, the specific weight of the open method of mining mineral deposits (75% of the total world mining) attests to the preservation of this general direction for the development of mining industries to provide fuel and mineral raw materials.

About 30% of coal, about 75% of iron ores, up to 80% of nonferrous metal ores, over 90% of nonmetallic minerals (asbestos, graphite, kaolin, mica, talc), almost 100% of nonmetallic building materials are produced abroad using open engineering.

In foreign countries with a developed mining industry, open-pit mining is conducted in the quarries "Flintkote Mine" (Canada), "Clevelanol Cliffs" (USA, "Westfrob Mine" (Canada), depth 244 m, "Palabora" (South Africa), "Aitik" (Sweden) In Table 1.1, the quarries of the developed countries of Sweden and the USA are considered (Zharmenov 2008).

Iron ore mining in the CIS is concentrated in the fields of Ukraine (Krivoy Rog basin), the Center (Kursk Magnetic Anomaly), Kazakhstan (Sokolovsko-Sarbaiskoye, Kacharskoye, Lisakovskoye, Ayatskoye, deposits) and the Urals (Arsentyev and Arsentiev 2002).

The extraction of nonferrous metal ores by the open method is mainly carried out in Siberia and Kazakhstan.

Increasing needs of Kazakhstan for fuel and mineral raw materials are provided through an open method of development, and include the extraction of coal, gold, uranium, iron, copper, nickel, lead–zinc, bauxite, and other ores.

Deposits of minerals, developed by the open method are characterized by a great variety of mining, –geological, mining, geomechanical, and technological conditions. The leading enterprises of the mining complex of Kazakhstan, which process open fields, are presented in Table 1.2 (Galiev 2003).

© The Author(s), under exclusive licence to Springer Nature Switzerland AG 2020
K.-K. Kassymkanova et al., *Geomechanical Processes and Their Assessment in the Rock Massifs in Central Kazakhstan*, SpringerBriefs in Earth Sciences, https://doi.org/10.1007/978-3-030-33993-7_1

Table 1.1 Indicators of quarries in the USA, Sweden

Indicators	Bingham Canyon (pcs. Yuta), Salt Lake City	Robinson (Nevada) Eli city	Morensi (Arizona) Tucson	Aitik (Sweden)
1	2	3	4	5
Height above sea level	2700 m	2700 m	2500	2000
Productivity:				
Rock mass, thousand tons/day	240	248	720	131
Ore, thousand tons/day	160	50		49
Content of Cu,%	0.52	0.69	0.1–0.45%	0.4
Amount in the quarry:				
workers	650 people	182 people		184
Support staff	180 people			20
Copper	270 thousand tons/year of cathodes		380 thousand tons/year of cathode copper	68 thousand tons
Gold, ton	8.5			1.5
Molybdenum	708,75 tons			
Silver, ton	99			55
Band width, mm	1600	1800	1800	1800
Strength of ore on the scale of prof. M. M. Protodyakonov	16–18	10–12	12–14	15
Warehouse volume	400 thousand tons			50 thousand tons
Copper recovery to concentrate	90%	74.9%		90%
Extraction of molybdenum	55%		52%	
Copper	300 thousand tons/year	Concentrate	380 thousand tons of cathode copper	240 thousand tons/year
Annual volume of rock mass, million m^3	48–50	32	97	96
Overburden, million m^3	25–29	27	44.4	60
Ore, million tons	60	15.3	52.6	18
Coefficient of overburden, m^3/t	0.5	1,8	0,8	1.7
Term of existence, years	30	9	20	8

Table 1.2 Mining complex of Kazakhstan

Variety of ore	Name of organization
1	2
Extraction of manganese ore	Mine "Tour" ("Kazchrome"), "Zhayremsky mining and processing plant"
Extraction of chrome ore	JSC "Donskoy mining and processing plant" TNC "Kazchrome"
Extraction of nickel ores	«KyzylKainMamyt» LLP
Extraction of copper ores	Corporation "Kazakhmys"
Extraction of lead–zinc ores	OJSC "Kazzinc", CJSC "Yuzhpolimetall"
Extraction of titanium–magnesium ores	Obukhov mining and processing plant-2, JSC «Mineral», JSC «Bektemir»
Extraction of rare-earth ores	"East Kazakhstan Company"
Gold mining	AK "Altynalmas", OJSC "Vasilkovsky mining and processing plant" OJSC "ГРК АБС-Balkhash» LLP "Kazakhaltyn"
Extraction of phosphorites	JSC MCC «Karatau»
Coal mining	"Ispat-Karmet", "Bogatyr Access Komir" OJSC "Eurasian Energy Corporation" "Semeykomir", etc.
Extraction of uranium	NAC Kazatomprom, Inkai JV, Katko JV
Extraction of asbestos	«Kostanai Minerals» LLP

About 35% of copper ore is mined at the quarries of the Severo-Zhezkazgan mine, the Nikolaevsky and Shemonaikha mines of the East Kazakhstan copper plant and Konyrat of the Balkhash mining and smelting complex. Since 2003, the new powerful quarry "Nurkazgan" has been working on the extraction of copper near the city of Temirtau. The design depth of the quarry is 600 m (Galiev 2003).

Due to the depletion of ore reserves located at accessible depths, the main development direction of the mining industry is towards the further development and improvement of the open method of mining operations, involving the exploitation of deposits with complex mining and geological conditions and large (up to 700 m) development depth (Melnikov et al. 2005).

In the theory and practice of open works, the main ways of increasing the completeness and quality of excavating minerals and improving the technical and economic indicators of quarries are known:

– an increase in the amount of overburden work with an increase in the depth of excavation to create favorable mining conditions when moving to underground work. But this option requires substantial additional material costs (Yakovlev 2009);
– revision of the initial draft of the final contour of the quarry and the formation of sides with increased angles of inclination. An increase in the angle of inclination of the pit on the final contour leads to a very significant decrease in the volume of stripping work. In particular, an increase in the angle of inclination of the pit

from 39 to 40° at a height of 400 m ensures a decrease in the volume of overburden in the pit contour by more than 345,000 m³ for every 100 m of the perimeter of the side (Kasymkanova and Tursbekov 2007). But the formation of steep sides requires a significant revision of the technology and organization of mining operations;

– involvement in the processing of abandoned poor ores, as well as mining waste, along with the extraction of the main mineral during the initial operation of the deposit;
– the use of temporarily non-working boards with a gradual working out of a penalty for reducing the stripping ratio;
– a combination of physical and technical production technologies with physical and chemical technologies,
– application of the combined open-underground method;
– rational use of the developed quarry space.

The main characteristics of some quarries in Kazakhstan are given in Table 1.3.

Analysis of the table shows that in Kazakhstan, various types of minerals are extracted by the open method: coal, iron, nonferrous metals, asbestos.

Thus, the issue of integrated development of the subsoil arises, and this, in turn, involves the involvement in the development of three groups of mineral resources according to the classification of Trubetskoy (1994):

1. Geogenic deposits of minerals;
2. Overburden, lying in the contours of quarries;
3. Technogenic deposits (external dumps of off-balance ores, dumps of overburden, tailings, metallurgical slag, and energy waste).

Table 1.3 Project parameters of ore quarries

Quarry	Angle of fall of the deposit, degree	The length of quarry, km	The width of the quarry, km	The depth of quarry, m	Angle of inclination of the ledges at working off	Height of non-working ledges, m
Sarbaisky	40–55	3.2	2.4	650	50–60	30–45
Uchalinsky	70–80	1.9	1.0	326	60	36
Sibay	40	1.4	1.4	471	30–65	30–40
Olenogorsk	65–80	3.9	0.85	385	50–60	24
Gaisky mining and processing plant	55–70	1.6	1.22	340	30–35	30
Kounrad	70–80	2.0	1.7	570	45–65	30
Sayak	75–90	1.9	0.73	300	50–60	30
Akzhal	50–60	2.0	0.80	240	60–70	30

In this connection, the main direction of development of the mining industry in the field of open works will be the transition to deeper horizons, accompanied by a complication of mining and geological conditions, an increase in the intensity of the massifs and a change in the deformation-strength characteristics of rocks. Therefore, the problem of geomechanical support for the stability of workings and the worked out space with new resource-saving technologies of open development comes to the fore.

At the same time, it is necessary to pay attention to the solution of the problem, including the following tasks:

- geomechanical substantiation of mining;
- optimization of career area parameters;
- analytical calculations of the volumes of overburden solids in internal heaps;
- substantiation of career transport and all technological processes;
- ensuring the stability of the sides of the quarry and the sideboards of the heaps;
- calculation of environmental risks and damage from the harmful impact of external heaps.

The specific nature of open mining of mineral deposits sets forth special requirements that must be taken into account already at the stage of designing quarries, identifying rational methods, and means of conducting mining. Among these requirements, which are of paramount importance, is the study of the influence of geomechanical processes on geotechnological parameters occurring in the instrumental and dump massifs.

To solve the problem problems facing mining science in this area it is required:

(1) carrying out research aimed at an in-depth and comprehensive knowledge of the environment in which mining is carried out;
(2) studying of natural phenomena, to deepen the knowledge in the field of the theory of rock pressure, the displacement of rocks, and the theory of artificial rock destruction (Kasymkanova et al. 2013);
(3) increase in the completeness of extraction of mineral resources from the bowels, protection of land resources, restoration of disturbed territories (Pospekhov 2004).

References

Arsentyev AI, Arsentiev VA (2002) Ways of development of technologies in the mining industry of the USA. J Min Mag 6:16–23 (In Russian)

Galiev SZh (2003) Prospects for the development of the scientific and technical potential of the mining sector in the light of the new industrial and innovative policy of Kazakhstan. Works Inst Min 65:10–20 (In Russian)

Kasymkanova KhM, Tursbekov SV (2007) Analysis of factors influencing the stability of quarry slopes. Min J Kaz 5:38–41 (In Russian)

Kasymkanova KhM, Nurpeisova MB, Jangulova GK, Baydauletova GK (2013) Harmony of
    subsoil in subsoil use. Bull KazNU 38:65–68 (In Russian)
Melnikov NN, Kozyrev AA, Reshetnyak SP, Kasparian EV, Rybin VV, Melik-Gaikazov IV,
    Svinin VS, Ryzhkov AN (2005) Conceptual principles of open pit wall design optimization,
    the Kola peninsula. Ivan Fyodorov Printing House, pp 3–14 (In Russian)
Pospekhov GB (2004) Engineering and geological surveys for reclamation of lands disturbed by
    the development of the Bogoslovsky brown coal deposit. Materials of the Urals mining decade,
    pp. 18–20 (In Russian)
Trubetskoy KN (1994) Open mining works. Mining Bureau, 590 p (In Russian)
Yakovlev VL (2009) State, problems and ways to improve open mining developments. Min J
    11:11–14 (In Russian)
Zharmenov (2008) Complex processing of Kazakhstan's mineral raw materials under the editorship
    of the academician of the National Academy of Sciences of the Republic of Kazakhstan.
    Monograph of the RSE NC of the Communist Party of the Republic of Kazakhstan №2-US-03
    Mining sciences and problems of development of mineral resources of Kazakhstan, 10:65–95
    (In Russian)

Chapter 2
Geological Setting Structural-Tectonic Features and Physical—Mechanical Properties of Mountain Rocks on the Stability of Slopes with the Account of the Time Factor and Large-Scale Explosions

Abstract Analysis of development experience, actual data on the stability of the sides of some of the ore mines in Central Kazakhstan shows that the efficiency of the open method of mining of mineral deposits can be significantly increased through the use of engineering management methods, which in turn is ensured by obtaining reliable information on the geomechanical state of the array.

Analysis of development experience, actual data on the stability of the sides of some of the ore mines in Central Kazakhstan shows that the efficiency of the open method of mining of mineral deposits can be significantly increased through the use of engineering management methods, which in turn is ensured by obtaining reliable information on the geomechanical state of the array.

Therefore, the problem of ensuring the sustainability of career slopes is the most important in mining. This is especially true for rocky and semi-local fractured massifs, since, in the case of high strength of individual monolithic blocks, the presence of weakening surfaces in the form of cracks in the form of long-distance cracks, the surfaces of tectonic disturbance mixers, contacts of layered rocks sharply worsens the stability of slopes.

Of the many factors influencing the stability of slopes with enclosing rock and semi-rock rocks, three main factors that require compulsory registration in the study of geomechanical processes can be distinguished:

(1) Structural and tectonic features of the mountain massif;
(2) Physical and mechanical properties of rocks;
(3) Researching of the influence of the time factor and the effect of mass explosions on the stability of slopes.

Of the above, the first two factors are natural, inherent in a particular array, they can only be taken into account in calculating the stability of slopes. The third factor is technogenic and should be managed when solving the problem of ensuring the

K.-K. Kassymkanova et al., *Geomechanical Processes and Their Assessment in the Rock Massifs in Central Kazakhstan*, SpringerBriefs in Earth Sciences, https://doi.org/10.1007/978-3-030-33993-7_2

sustainability of career slopes. All other factors have a subordinate value and can be accounted for in calculations through the safety factor.

Ensuring the stability of slopes and ledges of quarries is a complex task, the solution of which should include not only the definition of parameters of sustainable slopes but also their management for achieving better economic results and natural resources.

The structural structure of the massif is one of the main factors determining the strength and stability in the field development. The structure is understood as the nature and degree of fracture of the massif, which includes the linear dimensions of the cracks: their length and thickness, morphological features, the presence of aggregates, the spatial orientation of the cracks, their intensity, and a number of others characterizing the disturbance of the natural environment.

2.1 Research on Fracturing of Rock Massif

As noted by many researchers (Asanakunov et al. 2011), the influence of fracturing on the stability of the sides of quarries manifests itself in two main directions:

1. Decline the strength properties of the array. The evaluation of their influence, in this case, reduces to the determination of strength characteristics, both over the weakening surface and the most fractured massif.
2. Deformation of the massif as anisotropic or quasi-isotropic medium, the nature of deformation of a fractured massif is the basis of the choice of the design scheme. Consequently, in calculations of the stability of the sides, the choice of the design scheme is predetermined by the spatial orientation of the anisotropy, their extent and location relative to the sides of the quarry, and the calculated characteristics of the rock solidity depend on the size and shape of the structural blocks and the nature of the anisotropy surface (Iofis and Grishin 2005).

Statistical study of the fragmentation of the massif assumes the analysis of full-scale data of spatial orientation, crack capacity, intensity, and presence of a filler, its characteristics and a number of other morphological features. Research and analysis of all these parameters are mainly based on the data of geological services, which in the process of penetrating the mine workings lead to the certification of the structural pattern in the area of the development of the massif.

2.2 Research on Physical and Mechanical Properties of Rock Massifs

The physical and mechanical properties of the rocks in correlation with the structural-tectonic features of the mountain mass determine its tense state in the ledges and sides of quarries under the influence of internal and external forces.

A thorough and comprehensive study of the strength of the mountain massif must precede the solution of problems to prevent the deformation of slopes in quarries.

The main physical–mechanical properties of rocky and semi-local rock massifs, for solving the stability problems of slopes in quarries, are the density γ, the rock resistance to compression σ_0 and the fracture σ_f, the adhesion C, and the angle of internal friction p. These properties are different in the piece (sample) and in the array for the type of rocks. For example, the cohesion of rocks obtained from laboratory tests in a sample can be tens of times greater than for the same rock in the array. The angles of internal friction of rocks in the sample and in the array have an insignificant difference. In addition, the resistance of rocks to tearing in a monolith can reach a considerable value, and in a fractured massif, it is practically zero. The thing is that the real mountain massif is usually broken by a network of cracks, different in size, intensity of manifestation, etc. Therefore, the reliability of the obtained data is significantly influenced by the scale factor, as the characteristics of the rocks obtained from the tests of the elementary structural unit may differ from those in the mountain mass, which is a combination of elementary structural blocks. In this connection, for example, the bond of adhesion in a piece and an array is established through the coefficient of structural attenuation.

Because often an array of fractured rocks collapses on open surfaces on surfaces of attenuation of various origins, it is necessary to know the shear characteristics of these surfaces. The practice has established that the adhesion of cracks or contacts of rock layers can be several times smaller than the cohesion in the massif (Kuznetsova 2007).

The density of rocks, their resistance to compression and tearing, adhesion and the angle of internal friction in the piece are usually obtained in laboratory conditions by testing samples prepared from wellbores or by sawing them from stone cutting machines. For rock and semi-local rocks, the selected core or ore is usually not required to be refined.

The average density of samples of the correct geometric shape is determined by the formula:

$$\gamma = \frac{P}{V},$$ (2.1)

where P—sample weight, g;

V—sample volume, calculated by linear measurements, cm^3.

The average density of irregularly shaped samples can be obtained by hydrostatic weighing

$$\gamma = \frac{P}{P - P_1},$$ (2.2)

where P_1—sample weight in water, m.

Determination of the strength of rocks for uniaxial compression is performed in accordance with the international standard (GOST 21153.0-75 1975). The required number of test pieces is determined depending on the coefficient of variation.

V, % = 30, at the number of samples 9; cV, % = 25, at the number of 6 samples; V, % = 20, at the number of samples 4; cV, % = 15, at the number of 3 samples.

In the test, usually cylindrical or cubic forms are used with a ratio of height to diameter or base equal to one. The end support faces are ground, making them parallel with a deviation of no more than 0.5 mm. The sample is placed in the center between the plates of the press during testing. The loading rate is taken from 5 to 10 kgf/cm^2 s. before destruction.

The temporary resistance of rocks to uniaxial compression is calculated from the formula

$$\sigma_{cжc} = \frac{P_{max}}{S},$$ (2.3)

where P_{max}—breaking load, kg/n;
S—sample area, cm^2.

In practice, many methods are known for testing the tensile strength of a rock. The most reliable and available at present is the method for determining the tensile strength by splitting cylindrical specimens in diameter, called "Brazilian". The tensile strength of rocks in this method is determined by the formula:

$$\sigma_P = \frac{2P}{\pi \cdot d \cdot l}$$ (2.4)

where P—breaking load for sample;
d—diameter of the sample, cm;
l—height of the sample, cm;

Determination of strength characteristics of rocks C and p in laboratory conditions is based on obtaining normal σ_n and tangential stresses τ when testing specimens on a shear device with a press at various angles of inclination of the shear plane to the applied load. For the test, 3–4 samples of the same rock are made, they are placed in special clips and cut at angles equal to 30, 45, and 60° on a shear device. When cutting the specimen, the shear area (S) is measured, and the load (P) at which the shear occurred is taken from the press manometer. The normal and shear stresses at shear are calculated by the formulas:

$$\sigma_n = \frac{P}{S} \cdot \cos \alpha; \quad \tau = \frac{P}{S} \cdot \sin \alpha$$ (2.5)

Then the strength of the rock is compiled (Fig. 3). The resulting curve reflects the relationship between tangents and normal stresses. On the basis of VNIMI's investigations, the curve can be replaced by a chord in the section between the lines emerging from the origin at angles 45 and 60° to the σ_n axis (GOST 21153.0-75 1975).

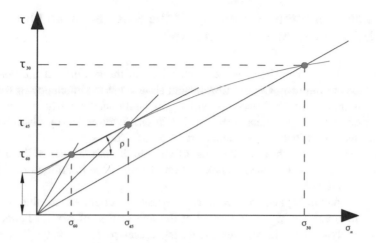

Fig. 2.1 Passport of rock strengthening

The angle of inclination of the chord to the abscissa axis determines the angle of internal friction of the given rock (ρ), and the segment cut off along the ordinate axis, the cohesion (C) on the scale of the graph (Fig. 2.1).

2.2.1   *Influence of Drilling and Blasting Works on the Stability of Ledges*

The question of the impact of drilling and blasting on the stability of ledges and sides of quarries composed of rocky and semi-local fractured rocks has not been sufficiently studied to date, although a number of works have appeared in the past 10 years that point to the serious importance of this issue and propose specific solutions.

Methods of conducting blasting operations in quarries significantly affect the strength, and consequently, the stability of the rocks. The collapse of individual ledges, and sometimes groups of ledges, in many cases is due to the fact that when approaching the limit contour, the anti-deformation mode of blasting operations was not observed.

The action of the blast wave extends to this surface, causing elastic and residual deformations, with the stress in individual directions, especially at their concentrations, reaching a considerable value exceeding the ultimate strength of the rocks, which causes irreversible deformation of the ledges and their destruction.

2.3   The Condition of Stability of the Sides of Some Ore Mines in Central Kazakhstan

Analysis of development experience, actual data on the stability of the boards of some of the ore mines in Central Kazakhstan shows that the efficiency of the open method of mining mineral deposits will significantly increase due to the use of engineering methods of management, which in turn is provided by obtaining reliable information on the geomechanical state of the array.

The most revealing object in the study of the stability of the sides of the pit is the Konyrat deposit, based on the main extraction of copper ore in difficult mining and geological conditions.

In the structural plan, the Konyrat deposit is confined to the stock of granodiorite-porphyry, which is located at the intersection of several faults in the core of the synclinal fold and represents the apical part of a large intrusion of granitoids not exposed by erosion. The granites of the stock are steeply dipping and complicated by numerous apophyses of varying thickness and shape (Kasymkanova 2007).

Folded disturbances were manifested in the collapse of the volcanogenic sedimentary strata into a large synclinal fold of the northwest strike with a steep fall of wings to the southwest. The syncline is sometimes complicated by additional folds of higher orders. The latter have dimensions not exceeding the first hundred meters in diameter, and are noted in the southwestern part of the deposit.

The disruptive disturbances in the deposit in terms of genesis and time of deposition are divided into regional pre-ore fracture systems, developed both in the rocks composing the deposit and beyond, and local systems of cracks developed exclusively within the field itself.

Semi-rocky and rocky, decomposed under the influence of moisture, rocks, represented by separate parts with a size of 0–1500 mm, are characteristic of the Konyrat mine. The natural moisture content of rocks varies between 2.5 and 3.0%.

The ore field of the deposit is composed of secondary quartzites, formed due to acid effusives and dacite-porphyries. The bulk of the ore is confined to secondary quartzites formed from dacite-porphyries and, to a lesser extent, to secondary quartzites formed from effusive sedimentary-metamorphic rocks. In poorly altered and unchanged rocks, mineralization is practically absent.

The rocks are greenish gray and characterized by a distinct porphyry structure. Almost all granodiorite-porphyries are intensively altered under the influence of hydrothermal-metasomatic processes.

The mineralogical and chemical composition of overburden depends directly on the type of minerals mined in the field, since for each type of fossil, there is always a characteristic association of associated minerals and chemical elements.

Secondary quartzites in granodiorite, granodiorite-porphyry (65%) are the most widespread in the deposit: quartz 34.6%, plagioclase 37.2%, orthoclase 21.3%, hornblende 4.2%, biotite—1.88%, accessory minerals—0.22%.

Secondary quartzites by acid effusions (30%): the bulk has a quartz-feldspar composition with phenocrysts of biotite and hornblende.

Silicate composition (%): SiO_2 = 74.0–76.0; Al_2O_3 = 13.0–15.0; CaO = 3.5–5.0.

The Konyrat deposit is represented by stockwork corpuscle of impregnated and fine-grained ores. In terms of stockwork, it has a horseshoe shape, almost isometric shape. Its maximum length is 1200 m, the average width is 700 m, the depth of mineralization reaches 500 m from the surface.

The copper and molybdenum mineralization is of industrial importance on the deposit, the distribution pattern of which determines the vertical and horizontal zoning of the deposit.

The oxidation zone has been worked out by now. The zone of secondary sulfide enrichment is developed very widely and has an important industrial significance. The lower boundary of this zone can be traced at depths of 350–400 m and has a very complex configuration.

Among the zone of secondary sulfide enrichment, quite often there are quite large areas of the prevailing distribution of primary minerals, copper; Transitions between the zone of secondary sulfide enrichment and the zone of primary ores are gradual, vague. There is a formation of "pockets" deep inside the zones.

Below the zone of secondary sulfide enrichment lie primary sulfide ores, represented by the impregnation of chalcopyrite and pyrite. A characteristic feature of the deposit is a decrease in the copper content with depth and to the periphery of the stockwork.

The distribution of copper is somewhat uneven, resulting in the allocation of areas of poor, medium, and rich ores. Molybdenum mineralization in the whole for the deposit is developed in the same range as copper but often goes over the contours of copper, forming independent concentrations. Areas of elevated metal concentrations are clearly gravitating toward the flank zones. A constant satellite of molybdenum mineralization is arsenic in the form of enargite and faded ores.

Thus, the vertical zoning of the ore body is expressed by a gradual decrease in the intensity of sulfide mineralization. Horizontal zoning is manifested in the concentric structure of the ore body with respect to its conditional center, which coincides with the barren core. Radius from the latter toward the contact of the rod of granodiorite-porphyries increases the intensity of copper mineralization, but near the contact itself it decreases again, and here the highest concentrations of molybdenum and arsenic, often associated with andalusite, are noted.

The main sulfur concentrations are associated with pyrite, evenly distributed in all sections of the deposit, which determines the uniform distribution of sulfur in the field.

The iron content slightly increases with depth, which is explained by chloritization, biotization, magnetization, developing on the lower horizons.

The main primary ore minerals are pyrite, chalcopyrite, molybdenite, enargite, fines, marcasite. Secondary—sphalerite, magnetite, bornite, galena, pyrrhotite, and rutile.

Ores are 90% composed of silicon and aluminum oxides. The chemical composition of the ores:—copper total—0.43%; copper oxidized—0.024%; molybdenum—0.006%; silica—75.2%; iron—4.3%; aluminum oxide—12.9%; calcium oxide—1.12%; sulfur 2.0%; silver—1.4 g/t; arsenic—0.008%; gold—traces.

Of all the useful components of practical importance are copper, molybdenum, sulfur, gold, silver, rhenium, selenium, and tellurium. The form of selenium and tellurium is not clear. The main mineral of copper in the primary ore zone is chalcopyrite.

Based on genetic and structural–morphological features, the Konyrat deposit belongs to the copper-porphyry industrial type. Figure 2.2 shows the structural section on the side of the quarry.

In terms of the size and shape of the ore bodies, the variability of their thickness, the internal structure, and the distribution of copper, the Konyrat deposit corresponds to Group I of the "Classification of Reserves of Reserves and Prospective Resources of Solid Minerals"—a large stockwork of a simple shape with a relatively uniform copper distribution.

Three types of rocks are involved in the formation of the Konyrat deposit: unchanged granodiorite-porphyry, gradually passing through the depth of occurrence into granodiorites, secondary quartzites over granodiorite-porphyries and secondary quartzites along effusive porphyries.

The rest of the rocks are very limited in the quarry area. These include quartz diorite-porphyry and secondary quartzites from diabase porphyrites.

Dykes of diorite porphyrites are found in the southeastern side of the quarry, where they cut the rocks of both intrusive and effusive complexes. The yields of secondary quartzites from diabase porphyrites are observed in the northeastern part of the quarry (Rakishev et al. 2007).

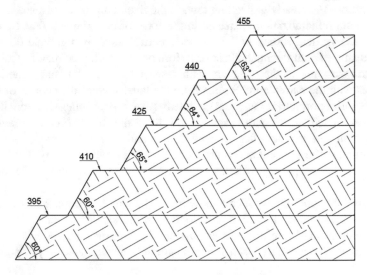

Fig. 2.2 Structural section on sideboard

Table 2.1 Characteristics of rocks and their strength

Name of rocks	Coefficient of strength by prof. M. M. Protodiakonov	Prevalence (%)
Secondary quartzites by effusions	12–14	30
Secondary quartzites for diorite-porphyries	10–12	60
Weathered diorite-porphyry and diabase porphyry	6–8	5
Unchanged diorite-porphyry	10–12	3
Monocarcites on effusions and diorite-porphyries	14–16	2
In average	11.4	

In addition, in the northeastern and eastern sides, as well as in the central part of the quarry, on the horizon of 545 m, there are kaolinized rocks that are developed in those areas where the effusive complex lies horizontally on the granodiorite-porphyry. The characteristics of the prevalence of rocks and their strength are given in Table 2.1 and the physical and mechanical properties in Table 2.2.

Table 2.1 shows the physical and mechanical properties of the rocks of the Konyrat deposit.

The study of the actual stability of the slopes, the revealed main types of deformations in the instrumental massifs, as well as the results of the structural features and physico-mechanical properties of the rocks, made it possible to obtain grapho-analytical relationships between the parameters of the slopes and the properties of the rocks.

Analysis of the dependence of the angle of slope on the angle of incidence of cracks shows that the fall of cracks toward the quarry significantly affects the slope of the ledge, and this dependence is rectilinear, i.e., a steep fall in the cracks causes a steep angle, and a gentle drop—a gentle slope of the ledge (Fig. 2.3). Such a dependence is observed when the cracks fall toward the quarry in the range 40–80°, and the characterization of the distribution of the systems of cracks in the Konyrat quarry is given in Table 2.3.

To identify the systems of cracks in the area of the collapse on the Konyrat quarry, the data are plotted on a rectangular fracture diagram. The vertical axis of the diagram characterizes the angles of incidence of cracks through the interval 10°, and to the horizontal azimuth of the strike of the cracks, through the interval—10°.

Each measurement is represented on the diagram by the center of a circle of small diameter. Then, contouring of individual fracture systems was performed, grouping them in the intervals of the angle of incidence and the strike azimuth equal to 30° (Fig. 2.3). As a result of this treatment, the diagram clearly identifies 6 systems of cracks.

Studies have shown that isolation of the site of a homogeneous fracturing is not performed accurately since large disjunctive disturbances are taken to the boundary of the site (Beck 2002).

Table 2.2 Physical and mechanical properties of the rocks of the Konyrat deposit

Main types of rocks	Density, t/m³	Strength characteristics in the samples		Strength characteristics in the array		The coefficient of structural attenuation according to G. L. Fisenko
		C_o (MPa)	ρ_o (°)	C_M (MPa)	ρ_M (°)	
Secondary quartzites	2.57	3.68	34	0.618	25	0.016
Granodiorite-porphyrites	2.54	3.12	35	0.52	26	0.016

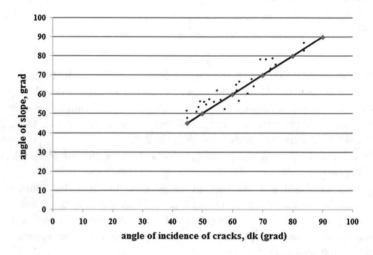

Fig. 2.3 Dependence of the angle of slope on the angle of incidence of cracks

For the sake of detail, polygonal curves are plotted for the angles of incidence of fracture systems along sections or for the entire quarry. Figure 2.4 shows the azimuthal polygonal curves for the distribution of cracks in the rock massif. It should be noted that on the quarry in the instrumental massif most often occur diagonal and longitudinal cracks, and according to the falling two times more often than the inconsistently falling.

Observations of the surface of the slopes of the ledges showed that over the course of time the magnitude of the inclination angles of the ledges decreases rapidly, and in the first 3–4 years a very intensive flattening of the inclination angles occurs, and later this process affects more slowly (Fig. 2.5). On the basis of the foregoing, it can be concluded that the flattening of the general angle of the inclination of the ledge is mainly due to the destruction of its upper part. Leaving safety, berms on one side are crumbled (in the upper edge of the ledge), and on the other side they are covered (near the bottom edge of the ledge), thereby connecting the slope into a continuous line. This makes it difficult to mine a career and requires systematic cleaning of berms and renovation of slopes, ledges, which ultimately leads to an increase in the volume of stripping (Fig. 2.6).

Table 2.3 Characteristics of the distribution of the systems of cracks in the quarry Konyrat

Crack systems	Elements of occurrence		Geometric classification
	Azimuth of extends, degree	Angle of decline λ, degree	
I	298	72	Transversal, according to the angle of inclination, a steep
II	14	74	Longitudinal, according to the angle of inclination, steep
III	338	79	Longitudinal, according to the angle of inclination, steep
IV	8	42	Longitudinal, according to the angle of inclination, inclined
V	75	75	Transversal, according to the angle of inclination, steep
VI	298	28	Transverse, according to the angle of inclination, sloping

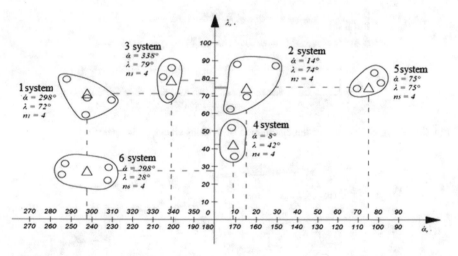

Fig. 2.4 Rectangular fracture diagram

The northern side of the Konyrat quarry, composed mainly of strong rocks, has permissible angles of inclination with two double ledges ($51°$–$54°$ at a height of 30–40 m); however, the absence of berms of the required sizes on the upper horizons requires a revision of their parameters.

The unsatisfactory state of the quarry sides is also due to the propensity of the rocks to fracture.

It is established that the main types of deformation of the ledges composed of rocky and semi-local fissured rocks are scree and collapse. On the ledges, scree and

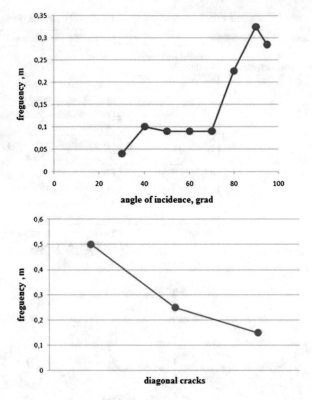

Fig. 2.5 Polygonal curves for the distribution of systems cracks in the quarry

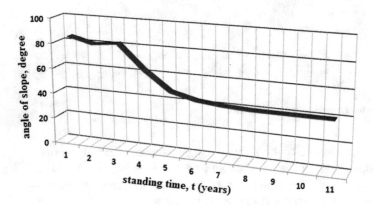

Fig. 2.6 Dependence of slope angle from the time of standing of the ledges

collapse are formed due to the impact of drilling and blasting operations on the stability of the rock massif.

The works on this subject are divided into three groups.

1. Work in which, based on mine surveying and seismometric observations, as well as empirically, the size of the zones of disturbance of the solidity of the massif is determined as a result of the action of mass explosions.
2. Works that contain general recommendations.

The influence of blasting on the stability of the bead in its limit position is proposed to be taken into account in two ways. The first is that to assess the stability of the side and calculate the angle of its slope as a whole, use those strength characteristics that have rocks that have experienced the impact of the explosion. The second way is to change the technology of works with the approach of the work front to the limit contour in such a way that the weakened zones of rocks arising in the explosion do not reduce the stability of the board, determined by the strength characteristics of rocks not affected by the explosion. The indicated dimensions of the zones along the surface of the ledge are confirmed by many authors (Beck 2002). Some authors argue that with mass explosions of vertical wells, the zone of burrows extends 6–13 m behind the newly formed upper curb, and the zone of shocks or residual shear deformations is 40–60 m and from the last row of boreholes. Here, the influence of the parameters of drilling and blasting on the value of the angles of inclination of the sides of quarries, folded by rock fractured rocks, is noted.

At the Konyrat quarry, the investigations of the disturbance of the massif by explosions are carried out by pouring water into the wells and determining the filtration coefficient. This method allows us to note the fundamentally new nature of the disturbance of the array not only over the surface, but also in depth.

It is known that as a result of the shock action of an explosion in the environment, stress waves arise that propagate at a high speed (280–5200 m/sec). A directional state created in a rock mass during the passage of compression and expansion waves leads (within the radius of destruction) to the emergence of a series oriented cracks due to the germination of natural equilibrium microcracks of the medium into macrocracks and the appearance of new stresses under the action of energy. With the subsequent expansion of the volume of gases after the explosion, the size of the cracks increases, which ensures a complete or partial disruption of the structure of the massif (Baikonurov and Melnikov 1970).

The radius of destruction produced by the energy in the front of the compression wave is insignificant in view of the fact that the depth of the germination of cracks is limited by the conditions of rock destruction in the state of all-round compression.

To determine the radius r, let us divide mentally the entire array, within the height of the charge, planes perpendicular to the axis of the well, into elementary layers of thickness Δh, and each layer, in turn, to elementary rings of width Δr (Fig. 2.7).

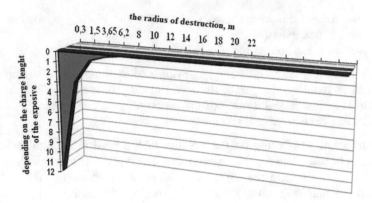

Fig. 2.7 Graph of change in the radius of destruction depending on the charge length of the explosive

Assuming the area of the elementary ring is 2 $2\pi r \, \Delta z$ unit, denoting by x its connection with the overlying layers, communication on the upper plane. Communication over the lower plane of the ring increases due to the increase in pressure by an amount Δx and will be equal to $x \div \Delta x$. The connection of such a unit ring will be a specific connection.

It is obvious that the destruction of each layer will begin when the acting stresses exceed a certain value of the bond.

The relationship between a layer of radius r can be determined by the formula:

$$S_0 = 2\pi \, r^2 (x + \xi h) \tag{2.6}$$

where ξ—change in the specific relationship per unit depth;

$h = n \, \Delta \, h$ (n—number of elementary layers).

Solving equation with relatively to r, we obtain

$$r = \frac{K_0}{\sqrt{h + C}}, м \tag{2.7}$$

where $K_0 = \sqrt{\frac{S_0}{2\pi\xi}}$—coefficient that takes into account the change in the strength of rock properties in the layer;

$C = \frac{x}{\xi}$—coefficient that takes into account the change in the strength of rock properties in depth.

The coefficients K_0 and C depend on the parameters of drilling and blasting operations and rock properties and are determined from the boundary conditions:

As a result of the research, it was revealed that in the approach of drilling and blasting operations to the design contour of the quarry, in order to exclude the deformation processes on the pit slopes that have already been put in the design position, in order to further safely conduct mining operations on the underlying horizons, it is necessary to study the mountain massif for its correction internal impairment using a variety of methods.

References

Asanakunov MA, Abdyldaev EE, Mashanov AA, Abdyldaev EK (2011) Accounting for joint fracture and contact conditions. Technicke vedy, pp 82–87 (In Russian)

Baikonurov OA, Melnikov VA (1970) Fundamentals of mining geophysics. Science, 326 p (In Russian)

Beck ASh (2002) Ensuring the stability of career slopes on the basis of strength properties. Dis. Cand., 150 p (In Russian)

GOST 21153.0-75 (1975) Rocks. Sampling and general requirements for physical testing methods. State Committee for Standards, 35 p (In Russian)

Iofis MA, Grishin AV (2005) The nature and mechanism of the formation of concentrated deformations in a tundish of shifting. Min Inf Anal Bull 7:82–86 (In Russian)

Kasymkanova HM (2007) A methods for researching the strength properties in a dump// Scientific and technical support of mining. Works Min Inst 73:233–236 (In Russian)

Kuznetsova IA (2007) Improving the technique for observing the deformations of the sides of the quarry (in the conditions of the Zherek deposit): dis. Cand. tech. sciences. 112 p (In Russian)

Rakishev BR, Mashanov AA, Teslenko TL (2007) Analysis of geotectonics in the design of mining enterprises. Mt Inf Anal Bull 3:25–30 (In Russian)

Chapter 3
Petrology and Geochronology of Country and Studying of the Disturbance of Rock Massif with the Use of the Teplometry Method

Local investigations of a pit slope placed in the limit position in the area where drilling and blasting operations are planned are recommended to be carried out using the method of thermometry to detect the depth of disturbance (cracks, shears, discontinuities, etc.).

This method is based on the relationship of heat output, when the radiation temperature, the infrared radiation of the studied section of the rock mass.

Experimental studies using the method of thermometry were carried out on different rock samples as complete and with small fracturing.

Scientists (Baikonurov and Melnikov 1970) in their works proved that the integral radiation of any body E_T is determined by the relation

$$E_T = Z \cdot \sigma - T^4, \ \text{t/m}^2 \tag{3.1}$$

where Z—body blackness degree ($0 \leq Z \leq 1$) and the value of this coefficient depends on its material, shape, and state of the surface.

σ—The universal Stefan–Boltzmann constant, which is determined by the formula

$$\sigma = \frac{2\pi^5 k^4}{15c^2 h^3} = 5.67 - 10^3 \, \text{t/m}^2 \, \text{degree}^4 \tag{3.2}$$

where $h = 6.626 \cdot 10^{-34}$ J/s—Planck's constant;
$c = 3 \cdot 10^8$ m/s—light speed in vacuum;
$k = 1.38 \cdot 10^{-23}$ J/K—Boltzmann's constant;
T—body temperature, °C.

Based on this, laboratory studies were carried out to establish the dependences of the intensity of thermal radiation on the temperature for different degrees of rock types.

© The Author(s), under exclusive licence to Springer Nature Switzerland AG 2020
K.-K. Kassymkanova et al., *Geomechanical Processes and Their Assessment in the Rock Massifs in Central Kazakhstan*, SpringerBriefs in Earth Sciences, https://doi.org/10.1007/978-3-030-33993-7_3

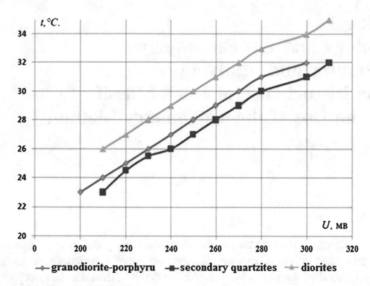

Fig. 3.1 Dependence of the output voltage of the photodetector on the surface temperature of the sample-emitters for various rocks

On the basis of laboratory equipment (a contact semiconductor thermometer that fixes the temperature of the sample surface, a millivoltmeter, a timer, a radiation meter of the sample surface based on a pyroelectric photodetector, a heat flux modulator, an oscillograph), the output voltage of the photodetector (imitating infrared radiation) and the thermophysical characteristics of the rock samples.

The dependence of the output voltage of the photodetector (imitation of the heat flux) on the surface temperature of the sample for granodiorite-porphyry, secondary quartzites for granodiorite-porphyry, secondary quartzites by effusive porphyries, diorites is shown in Fig. 3.1 and in Table 3.1.

The dependencies presented in Fig. 3.1 are described by the formula

$$U_{\text{вых}} = Z\,(T - T_{\text{окр}}) \ , \tag{3.3}$$

where $U_{\text{вых}}$—output voltage of photodetector, mV;

    T—sample temperature (measured by contact thermometer), °C;

    Z—coefficient characterizing the type of rock, mV/deg;

    $T_{\text{окр}}$—environment temperature (modulator curtains), °C.

To determine the temperature of an object from a known material, one can use the obtained dependencies (Fig. 3.2) and use them as calibrating for the intensity of thermal radiation.

This means that the temperature on the surface of the pit slope delivered to the design position will depend on the degree of fracturing and contrast at sunrise and sunset.

Table 3.1 Results of the dependence of the output voltage of the photodetector (simulating the intensity of the heat flux) on the surface temperature of the sample

Name of the rock sample	The environment temperature (degree)	The output voltage of the photodetector, (the intensity of the heat flux) (mW)	The temperature of the sample surface (degree)
1	2	3	4
Granodiorite-porphyry	23.0	288	32
		–	31
		239	30
		–	29
		195	28
		–	27
		137	26
		110	25
		90	24
		58	23
Secondary quartzites on effusive porphyries	22.0	306	32
		281	31
		246	30
		221	29
		195	28
		162	27
		133	26
		97.5	25
		–	24
		73	23
Diorites	23.3	–	32
		179. 5	31
		161	30
		140	29
		117	28
		88	27

It can be concluded that infrared photography should be done after the sun has sunk, it is only during this period that we obtain our own thermal radiation of the mountain mass forming the career escarpment.

The process of cooling granodiorite-porphyry rock samples, which are represented by different sizes and structures, was also studied to assess the dependence of the thermal physical characteristics of the massif from its disturbance (blockiness) (Fig. 3.2).

From the presented experiments, it follows that the intensive efficiency of the first-order production of the free surface of the sample of rock is inversely

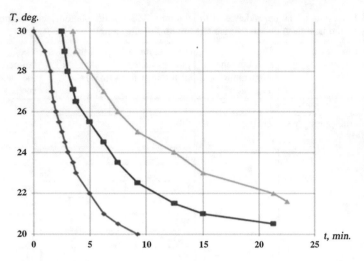

Fig. 3.2 Character of cooling of granodiorite-porphyry samples

proportional to its volume. We introduce a coefficient characterizing the shape of the sample and its linear dimensions

$$K = \frac{S_n}{V} \qquad (3.4)$$

Where S_n—sample surface area, mm^2

V—the volume of sample, mm^3

Based on the data given in Table 3.2 and the resulting graph shown in Fig. 3.2, we find how the heat transfer coefficient of the sample depends on the shape and size factor. The results of the calculations are given in Table 3.3.

Table 3.2 Cooling time of samples of granodiorite-porphyries

No. of experience	Dimensions of a sample of granodiorite-porphyry (mm)	The environment temperature (degree)	The temperature of the sample surface (degree)	Cooling time (min)
1	Sample in monolith 10 × 23 × 25	20	30.0	0.78
			29.0	2.1
			28.0	3.2
			27.0	4.9
			26.0	6.6
			25.0	8.4
			24.0	10.0
			23.5	–
			23.0	13.0
			22.5	–
			22.0	17.0

(continued)

Table 3.2 (continued)

No. of experience	Dimensions of a sample of granodiorite-porphyry (mm)	The environment temperature (degree)	The temperature of the sample surface (degree)	Cooling time (min)
			21.5	19.5
			21.0	28.0
			20.5	29.0
			20.0	33.0
2	Sample in monolith 25 × 27 × 24	21	30.0	0.5
			29.0	3.0
			28.0	5.1
			27.0	8.5
			26.0	12.3
			25.0	18.0
			24.0	24.1
			23.5	–
			23.0	31.0
			22.5	–
			22.0	47.1
			21.5	59.3
			21.0	85.1
			20.5	–
			22.5	–
			22.0	47.1
			21.5	59.3
			21.0	85.1
			20.5	–
			20.0	–
3	The sample is fractured 25 × 27 × 24	21	30.0	6.0
			29.0	10.0
			28.0	13.0
			27.0	20.0
			26.0	26.1
			25.0	34.1
			24.0	42.2
			23.5	–
			23.0	62.1
			22.5	75.0
			22.0	91.0
			21.5	–
			21.0	–
			20.5	–
			20.0	–

Table 3.3 The coefficient of heat transfer of the sample from the coefficient of form and size

Name	Measure	Number of experiment 1	Number of experiment 1
Dimensions of rock samples	Mm	10 × 23 × 25	10 × 23 × 25
b^{-1}	Min	0.54	0.415
		1.68	2.91
		2.94	5.61
		5.39	10.625
		7.92	17.712
		11.76	29.88
		16.1	46.995
		–	–
		26	72.85
		–	–
		39.1	143.184
		50.7	221.782
		84	
		107.3	
		–	–
K^{-1}	Mm	25	24
$b_{м} = b/K$	mm/min	46.3	57.83
		14.9	8.25
		8.5	4.28
		4.64	2.26
		3.16	1.36
		2.13	0.8
		1.55	–
		–	0.51
		0.96	–

Calculations show that the heat transfer rate of the samples, characterized by the coefficient b, is related to the coefficient of shape and size of K by a linear relationship. The nature of the cooling of the mountain mass is determined by its shape and size.

The data obtained experimentally on rock samples show that at the beginning the fissured sample cools faster than the monolithic sample, but then the rate of its cooling process slows down and at the end the fissile sample cools more slowly than the monolithic sample because the deep layers of the fractured sample retained more heat compared to the monolithic sample see (Fig. 3.1).

Chapter 4
Special Distribution of Ore Bodies and Teplometric Method of Express Evaluation of the Rock Massif Disturbance

In Chap. 2 of the monograph, the mechanism of the negative influence of career explosions on the stability of the sides was disclosed since the strength relations of the rock mass are destroyed (the mechanical properties of the constituent parts deteriorate).

Figure 4.1 shows the existing position of the quarry of the Konyrat deposit and its model at the end of the development (Fig. 4.1).

a—general view of the quarry model; *b*—cross section of the stretch; *c*—section along the stretch.

The working career ledge, broken as a result of explosions by cracks, is not dangerous, as it will be worked out in the process of mining (Fig. 4.2).

But when drilling and blasting works approach the design contour, it is necessary to carry out a number of measures to ensure the stability of pit slopes already placed in the design position, as the weakening of the mountain mass behind the design contour will affect its further stability in time and space.

Based on the laboratory tests carried out on rock samples, it is proposed to conduct studies on the disturbance of career slopes using the thermometric method.

The general model for drilling and blasting operations and infrared surveys for measuring disturbance on pit slopes by the method of heat metering is presented in Fig. 4.3.

The model shows a group of thermometers, two information processing units, three holes for thermometers, four slit wells, five small wells, and six series of wells.

(1) Inclined wells—slit (30 m angle 60°);
(2) From them series of wells (15 m)—near-well boreholes;

© The Author(s), under exclusive licence to Springer Nature Switzerland AG 2020   29
K.-K. Kassymkanova et al., *Geomechanical Processes and Their Assessment
in the Rock Massifs in Central Kazakhstan*, SpringerBriefs in Earth Sciences,
https://doi.org/10.1007/978-3-030-33993-7_4

Fig. 4.1 The existing position of the quarry of the Konyrat deposit

(3) In the direction of inclined wells (slit 30 m) from the near-line row of wells at a distance of 3 m, there are small vertical wells with a depth of 3 m;

(4) From the near-line number of wells at a distance of 7.5 m of the well (15 m) in the direction of the worked out space—wells for loosening the mountain massif.

The infrared survey to detect an internal disturbance of the mountain range that forms a career slope is as follows:

- in several places of a career slope we drill holes for the installation of thermometers, for full contact with the mountain massif;
- thermometers are connected with the information processing unit;
- electric signals proportional to the temperature of the rock recorded by the thermometers after the sun goes to the inputs of the microprocessor;
- the microprocessor performs the measurement and registers;
- according to the infrared survey, curves are plotted for the temperature of a highly fractured and slightly cracked mountain massif over time.

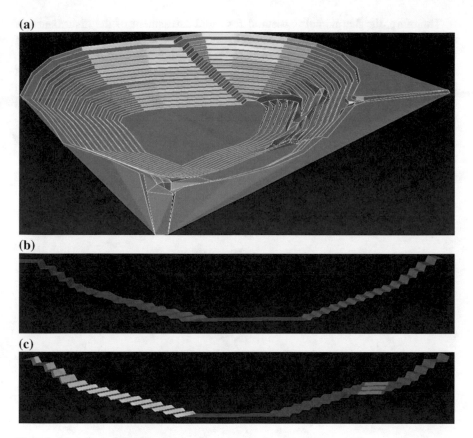

Fig. 4.2 Modeling of the Konyrat deposit quarry at the end of the working

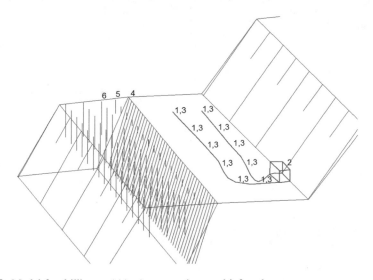

Fig. 4.3 Model for drilling and blasting operations and infrared surveys

Therefore, the thermometric method for rapid assessment of the disturbance of mountain massifs is characterized by accuracy, rapid measurement at simplicity of work, establishment of the border of homogeneous zones according to the degree of breach of a career slope, and on the basis of these data it is possible to correct drilling and blasting operations at the site adjacent to the pit slope delivered in design position that will increase the stability of the quarry sides for efficient and safe mining operations at the lower level horizons.

Chapter 5
Ultrasonic Method of Express Evaluation of the Rock Massif Disturbance

The ultrasonic method has found wide application in mining for determining the size of the disturbed zone, and for evaluating the physical–mechanical characteristics of the mountain massif.

In the present studies, the degree of disturbance of the massif and the amount of crack opening is established by the developed ultrasonic method, by sounding the monitored object and determining the transmission coefficient (D) of the elastic wave through the object. The method is based on the task of creating an uncomplicated method for determining the fracture of a mountain massif, which allows a more complete exploration of the rock mass in a shorter time.

The method is carried out by drilling wells in the rock mass (1) along the vertex corners of an equilateral hexagon and a borehole (2) in its center. Wells are located at the tops of the corners of two equilateral hexagons, with hexagons superimposed on each other, and the well (2) at the center of the first hexagon is a borehole at the apex of the angle (3) of the second hexagon. The ultrasonic wave receivers (4) are placed in the wells (1) located at the vertices of the corners of the hexagon, and the ultrasonic wave emitter (5) is placed in the borehole (2) located in the center of the hexagon. The ultrasonic wave receivers (4) and the ultrasonic wave radiator (5) are connected to the information processing unit by a multichannel microprocessor (6). The receivers and the radiator are fixed in the wells with a fast-hardening substance so that they are in the same line.

Then, the array is sounded between the wells by ultrasonic pulses, gradually changing the magnitude of the wavelength of the emitted pulse and measuring the magnitude of the amplitudes of the pulses transmitted through the array between the wells. Figure 5.1 shows the functional scheme of the ultrasonic method.

When an ultrasonic wave passes through a crack whose width is commensurable with the wavelength, the amplitude of the transmitted wave changes.

In the information processing unit (6), the transmission coefficient of the elastic wave can be determined immediately, which is equal to the ratio of the amplitude of the ultrasonic wave transmitted through the rock mass to the amplitude of the

K.-K. Kassymkanova et al., *Geomechanical Processes and Their Assessment in the Rock Massifs in Central Kazakhstan*, SpringerBriefs in Earth Sciences, https://doi.org/10.1007/978-3-030-33993-7_5

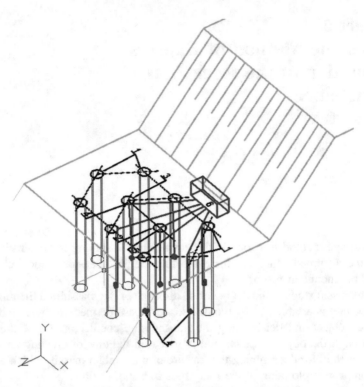

Fig. 5.1 Functional diagram of the ultrasonic method: 1—wells; 2, 3—vertex of a corner of a hexagon; 4—ultrasonic wave receivers; 5—ultrasonic wave radiant; 6—multichannel microprocessor

ultrasonic pulse emitted into the rock mass, and the dependence of the transmission coefficient of the elastic wave on the wavelength is constructed (Yakovlev 2009).

The coefficient of propagation of elastic waves D in the first section does not depend on the amplitude of the ultrasonic wave emitted into the sample. So D is significantly less than the transmission coefficient for an undisturbed D_H object. Hence, it follows that the sample contains defects, but it is not possible to reliably determine their nature and parameters.

By a stepwise increase in the transmission coefficient of ultrasound, the presence of a crack in the array is judged, and the magnitude of the crack opening is judged from the amplitude of the emitted wave corresponding to a stepwise increase in the transmission coefficient.

(D) from the amplitude of the radiant signal (Au)

The amount of fracture opening is calculated by the formula:

$$L = A_{\kappa p} \cdot e^{-\delta l_0} \tag{5.1}$$

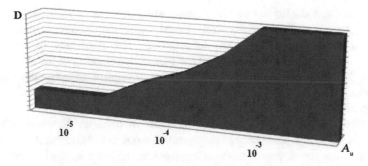

Fig. 5.2 Change in the value of the transmission coefficient of elastic waves

Where A—amplitude of the radiated ultrasonic wave;
Δ—ultrasonic attenuation coefficient in the sample;
l_0—distance from the excitation point of the ultrasonic wave in the sample to the crack.

The distance l_0 is determined in a known manner, for example, by an echo method. The amount of crack opening L can be expressed in conventional units or in units of length. In this case, $A_{кp}$ is determined by the product of the reading of the peak voltmeter U and the coefficient of the radiator conversion. If there are several cracks in the controlled object, then the dependence $D = f(A_u)$ has a multistage form (Fig. 5.2).

A sharp increase in the coefficient to values close to the value of the transmission coefficient D_H at $A_{кp}$ for an undisturbed object is due to a change in the mechanism of energy transfer of the elastic wave through the crack at vibration amplitudes commensurate with the crack opening. In this case, the energy is transferred mainly not through the filling of the cracks, but directly from one wall to the other.

The observed asymptotic approximation of the values of $D_к$ D_H in the last section is explained by the fact that for amplitudes of vibrations exceeding the crack opening, practically all of the energy without losses is transferred from one wall to the other. The acoustic wave, in this case, does not react to the crack.

The application of the method makes it possible to carry out simultaneously six measurements, which accordingly accelerates the process of studying the massif and the completeness of the investigation.

Further studies to determine the areas of the of arrays with varying degrees of risk of deformation, the detection of the boundaries of unstable blocks of the device array, and the features of their structure are possible only on the basis of methods that allow the development of deformation phenomena to be recorded before manifestations of solidity discontinuity in the array are manifested (GOST 21153.0-75 1975).

Widely applied methods of visual and instrumental surveying and geodetic observations are reduced to fixing visible manifestations: violation of the stability of slopes, the nature of the joint fracture, the consequences of blasting in the zone of their influence, and also to obtaining quantitative patterns of slope deformation in order to determine the velocity and critical value displacements preceding the beginning of the active stage of deformation. The frequency of observations depends on the intensity of mining and the rate of deformation of the side. It should be noted that surveying observations of the state of stability of the quarry's sides are very laborious and do not allow us to determine the earliest stage of deformation development in the depth of the massif.

To solve this problem geophysical methods of research (thermometric, ultrasonic) that have high sensitivity to structural changes in the array at the early stages of deformation development allow. Geophysical methods applied at the Konyrat quarry are aimed at solving the following problems:

- zoning of the sideboards of the quarry for stability;
- observation of the geodynamic activity of major tectonic disturbances within the pit;
- long-term and operational prediction of the stability of local sections, taking into account the technogenic impact of mining operations.

The results of the studies showed that the fracture of rocks in the deposit is distributed unevenly both in depth and in area.

5.1   Complex Using of Ultrasonic Signal for Evaluating the Disturbance of the Rock Massif

The methodology for rapid assessment of the disturbance of the mountain massif in order to monitor it operatively can be used both in a complex and in conjunction with other methods, for example, with surveying field fracture measurements that complement each other.

Based on the determination of the propagation velocity of longitudinal and transverse elastic waves, using appropriate ultrasonic equipment and measurement techniques.

The measured propagation velocities of longitudinal and transverse elastic waves allow us to obtain with certain accuracy, dynamic parameters such as the elastic modulus E_n and Poisson's ratio μ,

$$E_n = V_P \cdot \rho \frac{(1 + \mu) \cdot (1 - 2\mu)}{(1 - \mu)}; \qquad (5.2)$$

where V_p velocity of spreading of longitudinal waves, m/s;
  r—Rock density, kg/m^3;
To determine the Poisson's ratio use the following formula:

$$\mu = \frac{0.5 - \left(\frac{V_S}{V_P}\right)^2}{1 - \left(\frac{V_S}{V_P}\right)^2}; \tag{5.3}$$

where V_s/V_p—the ratio of the velocity of propagation of a transverse wave to the velocity of a longitudinal wave.

For laboratory studies of the propagation velocity of longitudinal and transverse waves, the technique and equipment described in the literature (Jangulova and Tulebaev 2014) were used.

As a device, a high-frequency generator from MATRIX Corp. was used. MFG-8216 and digital storage oscilloscope (CCD) of ACUTE Tecn. DS 1002 (Methodical instructions 1988; Savich 1969).

For combining system requirements, PCs (personal computers) such as Pentium 1.2 GHz, Pentium 2.0 GHz, with RAM 64 Mbytes of RAM (min) and the operating system Microsoft Windows XP, etc., were used.

Piezoelectric transducers PRIZ-12 (TU 25-7761.033-86) are used as transducers (Fig. 5.3).

The main parameter of the measurement in determining the elastic properties of rocks is the acoustic delay time Δt of the front of the first half-cycle of the pulse of the received signal, which for a constant sample size L = 100 mm allows us to determine the propagation velocity of the longitudinal wave $V_p = L/\Delta t$. To determine the propagation velocity of the transverse wave, the signal reception sensor (piezoelectric transducer-radiating or receiving) was located on the sample at an angle (Galustyan 1989):

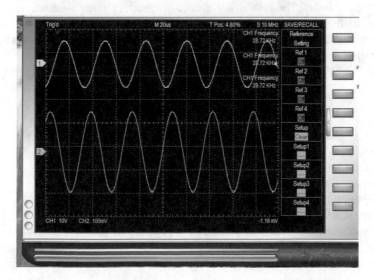

Fig. 5.3 Ultrasonic method of researching

Fig. 5.4 Block diagram of
the measurement of elastic
parameters by the pulse
method: 1—generator; 2—
oscilloscope; 3—radiating
piezoelectric transducer;
4—п receiving piezoelectric
transducer; 5—acoustic load
—a sample of rock made of
core or standard; 6—
computer for data input and
pulse monitoring

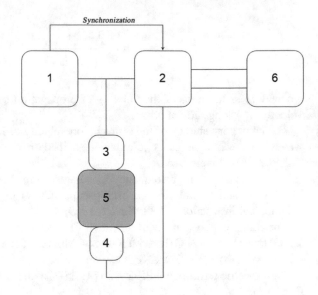

$$90°V_s = L(90°)/\Delta t \tag{5.3}$$

where L (90°)—length of the sample taking into account the displacement of the
sensors on the sample surface. Figure 5.4 shows a block diagram of the measurement of elastic parameters by the pulse method.

The data of V_p and V_s made it possible, on the basis of the formulas of the theory
of elasticity, to perform mathematical calculations of the values of the Poisson's
ratio and the elastic modulus E_n for rocks.

Samples selected for measuring elastic parameters were selected in all well
intervals with a complete representation of all the rocks described along the well.

To measure the fracture voidness, which fully characterizes the fracture of rocks
by sounding a sample was determined by the formula:

$$T = \frac{V_{p1}\Delta t - d}{V_{p2} - V} V_{p2} M \tag{5.4}$$

where V_p—the maximum speed of transmission of longitudinal waves in a
monolith, m/s;

Δt—the transit time V_p in the sample, m/s;

d—linear sample size, cm;

V_{p2}—velocity of longitudinal waves in the sample, m/s;

V—speed of sound, m/s.

The application of the method makes it possible to perform measurements on
samples, which accordingly accelerates the process of studying the rock mass and
completeness of the study.

According to samples selected for measuring elastic parameters, samples were chosen in all intervals of several wells with a complete representation of all the rocks described, and the results of determining the longitudinal and transverse wave velocities in Table 5.1.

The internal friction angle (φ) and the clutch (C) were determined from the strength certificate. The passport of the strength of the rock is a curve that envelops all the maximum Moore stresses, constructed in the coordinates of normal σ and tangential stresses τ (Fig. 5.5).

The passport of the strength of rocks is built according to GOST 21153.0-75 (1975).

To measure the fracture voidness, which fully characterizes the fracture of rocks by sounding a sample, was determined by the formula:

$$T = \frac{V_{p1}\Delta t - d}{V_{p2} - V} V_{p2} M \tag{5.5}$$

where V_p—the maximum speed of transmission of longitudinal waves in a monolith, m/s;

Δt—the transit time V_p in the sample, m/s;

d—linear sample size, cm;

V_{p2}—velocity of longitudinal waves in the sample, m/s;

V—speed of sound, m/s.

5.2   Research of the Stress-Deformational State of the Rock Massif Using Modern Equipment Operating on the Non-destructible Testing Method

Estimation of the stress–strain state (VAT) of the rock massif by natural methods, in spite of their high reliability, is associated with a number of difficulties; therefore, when studying the stress state, it is expedient to conduct laboratory experiments on models (Kasymkanova 2007). However, with a large variety of the structure and properties of the mountain massif of deposits, it is difficult to develop a model that could describe the stress–strain state of all species of rocks encountered in nature. Therefore, the improvement of the compilation of models of stress–strain state, reflecting the features of the structure and properties of rocks, is topical (Iofis and Grishin 2005).

In recent years, the methods of work have changed radically, thanks to the development of geoinformation systems (GIS) technologies. These methods made it possible to solve complex problems for various physical models, analyze the deformation and strength properties of the rocks in the Akzhal field. The types of rocks composing the Akzhal deposits and their properties are given in Table 5.2.

Table 5.1 Strength and elastic properties of rocks from control wells

Well	Rocks	Depth of sampling (m)	Strength properties				Elastic properties					
			Volume weight γ · 10⁻³ (g/cm³)	Fractured voidness T, %	Strength limit for compression $\sigma_{сж}^{*}$ (MPa)	Strength limit for stretching σ_{p}^{*} (MPa)	Elastic modulus E · 10⁻⁴ (Mpa)	Speed of longitudinal waves V_p (m/s)	Transverse wave velocity V_s (m/s)	Poisson's ratio (μm)	Linkage force Mpa	Angle of internal friction, φ degr
Sample of rocks from wells												
1	Porphyrites	10	2.54	8.6	75.14	16.7	4.53	4760	2660	0.26	17.26	38
	Porphyrites	25	2.54	8.1	100.55	14.02	3.9	4700	2412	0.32	20	47
	Porphyrites secondary quartzites	38	2.62	9.2	57.45	12.24	4.92	4650	2770	0.225	17	40
	Secondary quartzites	45	2.62	7.0	42.54	11.04	5.21	4650	2900	0.178	11.5	44
	Secondary quartzites	60	2.62	7.5	76.8	14.5	4.54	4720	2600	0.282	16	36
	Porphyrites	70	2.54	5	44	14	6.06	5050	3200	0.16	15	25
	Porphyrites	79	2.54	8.1	44.47	15.91	4.68	4690	2720	0.216	14	22
	Porphyrites	98	2.54	7.05	48.61	13.64	5.58	4739	3160	0.2	12.5	34
	Porphyrites	104	2.54	8.7	27.07	12.8	3.94	4670	2430	0.31	10	19
	Porphyrites	120	2.54	8	50.27	12.8	5.72	4450	2585	0.27	16	40
	Porphyrites	122	2.54	8	29.3	10.2	4.96	4350	2600	0.25	9.5	33
2	Diorites	10	2.72	10	16.8	2.7	3.29	3584	2290	0.155	4	40
	Porphyrite plagiogranites	15	2.72	15	13.48	3	3.75	3650	2310	0.165	4	38
	Secondary quartzites	18	2.72	15	19	3.81	3	3500	2190	0.18	4	40
	Porphyrites	36	2.54	8.5	21.88	4.29	3.66	3906	2500	0.16	4	42
	Porphyrites	50	2.62	10	15.14	4.2	1.91	2960	1710	0.25	4	40

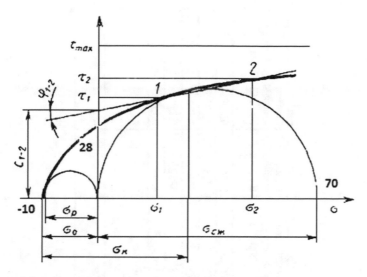

Fig. 5.5 Building a strength passport

The most convenient method for calculating the stress field for spatial regions with a complex contour is the method of boundary integral equations (SMI) (Beck 2005). The stress–strain state of the instrumental massif in the marginal outline of the pit was calculated taking into account the structural-tectonic features of the massif and external loads (Fig. 5.6).

The state of the rocks in the calculations is assumed to be quasi-isotropic with a coefficient of brittleness and plasticity $\chi = \frac{\sigma_p}{\sigma_{c;}} < 0,2$, and the change in the properties of the massif in time under the influence of the stressed state (manifestation of creep) has been taken into account (Mashanov and Nurpeisova 2000).

From Fig. 5.6, which shows isolines of stresses at the base and sides of the pit at $\gamma H = 0.027$ MPA, the current moment and the period of setting it to the limit position shows that an appreciable concentration of normal stresses σ_x and tangential τ_{xy} is observed in the zones adjacent to the bottom of the bead at points of inflection of the curvilinear boundary.

It is noted that at an existing height of the ledges 10 m H_{yc} 30 m and a width of berms of 10 m B 30 m, the configuration formed by them does not affect the stress state. As these parameters increase, their influence is noticeable (Fig. 5.6a, b). The distribution of normal vertical stress σ_y increases smoothly along the side as a function of depth (Fig. 5.6c).

The normal horizontal stresses σ_x in the upper part of the bead are 20–25 times larger than σ_y, and at its base are 3–5 times higher than the vertical ones, and the ratio σ_x/σ_y increases with the depth of the quarry H. At the earth's surface, the maximum concentration σ_x occurs at a distance kH (κ is a coefficient depending on the height of the bead H and the angle of its inclination α) from the edge of the bead (Fig. 5.6d).

Table 5.2 Physical and mechanical properties of the rocks of quarry Akzhal

Rocks	Density γ, 10^3 (kg/m^3)	Adhesion C, MPa	Angle of internal friction ρ (deg)	Strength limit for compression, $\sigma_{сж}$ (MPa)	Strength limit for stretching, σ_p (MPa)	Elastic modulus E (MPa)	Coefficient of structural weakening by L. G. Fisenko
Massive limestone	2.7	28	32	110.5	13.0	8756	0.166
Limestone crushed	2.68	30	32	136	14.3	8725	0.154
Diorite-porphyrites	2.7	53	32	153.1	14.8	7950	0.172
Porphyrites	2.7	10	31	140	14.8	6837	0.156
Limestones	2.68	25	30	128	13.5	8364	0.150
Clayey-limestone rocks of zones of tectonic crushing	/2.67/	/0,04/	/30/	95.1	17.8	–	–
Rocks of zones of tectonic disturbances	/2.6/	/0.04/	/30/	90.5	17.9	–	–

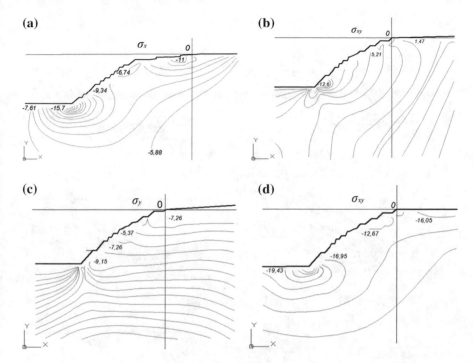

Fig. 5.6 Voltage isolines in the sideboard of array

Foci of normal stress concentration σ_x adjoin the parts of the massif boundary that have the greatest curvature, and they determine the location of the most dangerous zones in the mountain massif.

Thus, the term "undestroyable control" does not correspond to those processes that occur in the array. Indeed, if we compare the distribution of stresses in rock massifs before and after the excavation, we note that the normal stresses $\boldsymbol{\sigma_y}$ decrease mainly, and the remaining stress components increase and contribute to the formation of stress concentration and plastic flow zones, in fact, there is a process of destruction, and indestructibility (Fig. 5.7).

The simulation of this process is carried out on the basis of the strength function (Report № 003219, KarPTI 2001), transformed into an expression.

$$F = (\sigma_x + \sigma_y + 2Cctg\phi)^2 \sin^2 \phi - [(\sigma_x - \sigma_y)^2 + 4\tau_{xy}^2] \qquad (5.6)$$

where σ_x, σ_y—components of the normal stresses on slip areas, MPa;

C—adhesion, MPa;

φ—angle of internal friction, degree;

τ_{xy}—tangential stresses on sliding areas, MPa.

Figure 5.7 shows the nature of the change in the isolines of strength in the array at different depths of the quarry, adhesion, and the angle of internal friction (the

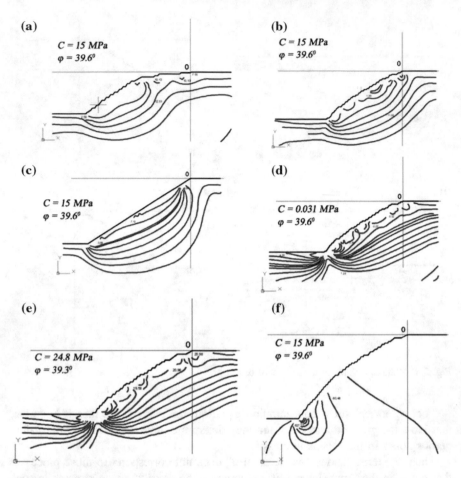

Fig. 5.7 Character of the change in the isolines of strength in the array for different values of quarry depth, adhesion, and angle of internal friction

zero isoline of the limiting state is shown in blue, isolines with negative values correspond to the fracture zone).

In Fig. 5.7a, b, with an increase in H from 265 to 600 m for constant C and φ, the volume of the array in an unstable state increases.

At $H = 265$ m, the softening zone is absent, and the massif has a sufficient margin of safety; at $H = 360$ m, a softening zone is formed in the near-surface region at a level of one-third of the height of the bead from the upper edge. For a depth $H = 600$ m, the formation of this zone in the form of an exponential curve is typical, and the limit state line goes to the contour of the side in the upper part at a distance of 100 m from the upper edge.

When the strength properties of rocks are reduced, the configuration of the zones of softening changes to a considerable extent. Thus, in the rock strength index typical for the areas of tectonic disturbances (Table 5.2), the zone of weakened rocks with the examined side parameters (H = 600 m) extends along the surface of the bead to a depth of 300 m, and along the base and the earth's surface—up to 200 m (Fig. 5.7c). Thus, rock areas with these properties in this zone become less stable and serve as potential shear surfaces. For very strong rocks, there is no softening zone (Fig. 5.7d).

Note that the softening zone is largely determined by the combination of the constituent stress components forming on the elementary area and entering into the above expression (5.6). At the depth of the quarry side up to 600 m, the softening zone is formed as a region of radius about 200 m adjacent to the bead base at the point of inflection of the curvilinear boundary (Fig. 5.7f). In this case, the zero line extends to the board surface approximately in the middle part, changing the bead configuration, it is possible to reduce the softening area, thereby increasing its stability.

Thus, the change in the boundary conditions as a result of the deepening of the quarry will cause a redistribution of the constituent stress components and an increase in the volumes of the rock mass that is in an unstable state. The presence of sites with reduced strength characteristics increases the probability of deformation of the bead.

The length of all destructive deformations along the front, as a rule, exceeds the height of the deformed ledge by 2–5 times. This indicates that the influence (clamping) of the lateral rocks on the stability of the ledges in the limit position affects only as long as the extent of the disturbed section does not exceed its height. Therefore, the establishment of working frames along the contour of the quarry makes it possible to control the stability of the sides throughout their entire length, and the network created is used as a basis for expanding observations.

5.3   Studies on the Establishment of a Grid of Crevice Wells by Method of Instrumental Observations Using Modern Geodetic Equipment

Depending on the change in the parameters of drilling and blasting operations and the order of working off the ledges in the contour zones, the existing methods for reducing the harmful effect of mass explosions on the stability of ledges and quarries of quarries in the application of which the parameters of drilling and blasting operations in the contour zones vary, and the order of working of the ledge remains unchanged; This includes the methods of drilling and blasting operations, which allow reducing the degree of destruction of the rock massif in the contour zones.

The most common is the blasting along the outline of a pit of inclined wells with the restriction of explosives in them with a thickened grid, i.e., creation of a screening gap (Okatov et al. 2003; Tsai et al. 2005).

To reduce the harmful effect of mass explosions on the stability of the rock massif, it is necessary to arrange a screening gap between the blasts and the protected blocks. The method of forming a screening gap by the drilling and blasting method is as follows:

(1) In the production of blasting operations in the contour zones (20–40 m from the pit contour), the rows of boreholes are not parallel to the contour line, as it was before, but perpendicular to it.

(2) On the prospective contour of the pit, at 2–3 m from the extreme wells of each row (depending on the required angle), a contouring series of holes with a desired angle of inclination and a depth of 3.5–4.0 m is drilled, but not less than one-third of the height of the ledge. The distance between the holes is 0.8 and 1.0 m. The blast holes are charged with dispersed charges of the patronized explosives and explode simultaneously with the first series of wells (Popov et al. 1997).

The use of the proposed method will allow:—to reduce the size of the rock separation zone from the extreme wells of each row (from the last row of blasting holes) to 3.0–3.5 m instead of 6.0–7.0 m with the usual method of drilling and blasting operations and to increase the angle of the slope of the ledge by 10–150; prevent the formation of cracks and cracks in the array beyond the zone of the new edge of the ledge, which greatly increases its stability; provide, when the sides of the quarry are covered, with a double edging at an angle of 70° (Fig. 5.8).

The collapse of individual ledges, and sometimes of ledges, is in many cases caused by the fact that when approaching the limit contour, the anti-deformation mode of blasting was not observed (Galiev and Shamganova 2012; Maistrovsky and Tulebaev 2010).

Fig. 5.8 Scheme of strengthening and elaboration nonworking benches in strong homogeneous rocks

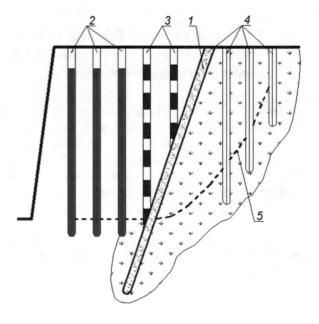

The action of the blast wave causes elastic and residual deformations, with the stress in individual directions, especially at the places of their concentration, reaching a considerable value exceeding the ultimate strength of the rocks, which causes irreversible deformation of the ledges and their destruction. Based on the instrumental observations with the use of modern geodetic equipment, the sizes of the zones of disturbance of the solidity of the massif are determined as a result of the action of mass explosions. In the course of instrumental observations with the help of "beacons" placed on the upper and lower areas before the explosion, it was established that signs of shifts occur at a distance from 40 to 60 m from the blasting blocks. As you move away from the incurred block, the degree of disruption of the array decreases.

The influence of blasting on the stability of the bead in its limit position is proposed to be taken into account in two ways. The first is that to assess the stability of the side and calculate the angle of its slope as a whole, use those strength characteristics that have rocks that have experienced the impact of the explosion. The second way is to change the technology of works with the approach of the work front to the limit contour in such a way that the weakened zones of rocks arising in the explosion do not reduce the stability of the board, which is determined by the strength characteristics of rocks not affected by the explosion. In the case of mass explosions of vertical wells, the zone of burrows extends 6–13 m beyond the newly formed upper edge of the ledge, and the zone of shocks or residual shear deformations is 40–60 m from the last row of wells. The deformation of the ledge under the action of the explosion depends on the fracture of the massif and the elastic properties of the block material.

5.4 The Way to Strengthening Career Slopes of Folded Fractured Rocks

In the practice of open development of the field, there are many cases when due to improper evaluation of stability and acceptance of high values of slope angles of the sides of the quarry, large landslides and caving occurred, and unjustified reduction of the repayment angles led to significant costs for extracting additional volumes of overburden. Collapses in quarries mostly occur on the surfaces of weakening caused by disjunctive disturbances, as well as by the presence of cracks directed toward the developed space. Various kinds of caving in lead to significant violations of the operating mode of the quarry and require significant costs for the restoration of transport routes.

One of the effective methods to prevent this kind of deformation of rocks is their artificial reinforcement, which makes it possible to provide the necessary stability of the slopes of the ledges of the nonworking sides of quarries and in some cases to prevent possible collapse of rocks in weakened areas, in others—to significantly reduce the amount of stripping work.

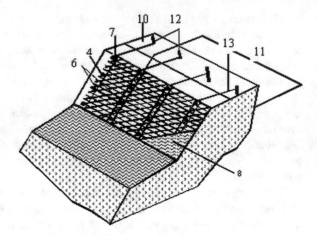

Fig. 5.9 Reinforced ledge in isometric view: Between screening wells (1) drilling additional wells (3) at the same angle as the screening wells and lay the armature in them (4), for pipes it is possible

We propose a new way to strengthen quarries folded by fractured rocks. Figure 5.9 shows the reinforced ledge of the pit in a isometric view.

The essence of the method consists in the following: when approaching mining operations to the boundary of the drilling drill, inclined shielding wells 1 with a distance between them of 1.5–3 m, depending on the rock hardness (Figs. 5.10, 5.11).

Screening wells (1) are charged and explode. After the explosion, a zone of rock crushing is formed (5). Additional inclined wells 3 are equipped with conductors (not shown) and pumping in them a reinforcing solution, for example, a concrete mix, all the way to the upper mark and then hardening it. In this case, the armature (4) must be adjacent to the wall of the well (3) from the side of the slope (2).

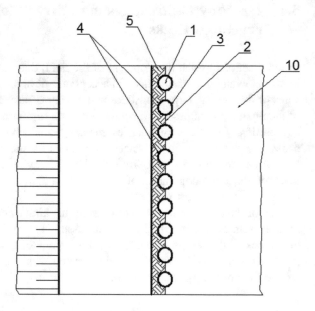

Fig. 5.10 Working area at the plan

Fig. 5.11 A reinforced ledge in the section

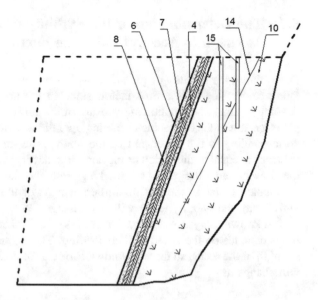

The remaining mountain massif in front of the slope line (2) is blown up. Then, the blasted rock mass is cleaned and the slope surface is peeled off (2). As a result, the armature is exposed (4) to the armature weld rods or hooks (6). Then they pull on the metal mesh (7) and fix it. A layer of a strengthening solution (8) (for example, a layer of a concrete solution) is applied to the mesh by the spraying method. After the layer of the solution has hardened, the slope will be safely secured and protected from weathering.

In addition, behind the proposed shear plane 9, vertical wells (11) drill at a depth of 3–4 m opposite the additional inclined wells (3) on the drilling site 10 and anchors (12) are installed therein. Between the armature (4) and the anchors (12) pull the cables (13). This increases the reliability of strengthening the ledge of the quarry.

If a fracture is formed as a result of the explosion (14), in this zone vertical wells (15) are drilled in the drilling site (10). The wells are equipped with conductors and a reinforcing solution is injected, for example, a cement or special solution.

In most quarries, rocks forming nonworking beads are heterogeneous in structure and physico-mechanical properties. Even in relatively uniform eruptions of the rocks of the open field, there are usually 5–6 or more zones of tectonic disturbance. Depending on their size and spatial orientation, they have a more or less significant effect on the stability of the beads and ledges. Strengthening with a special solution will improve the safety of operation, increase production productivity.

5.5   The Determination of the Stability of Quarry Slopes, Taking into Account the Time Factor

Determining the stability of pit slopes with one condition is the method of calculating career slopes, is a time factor, since in the array occur deformations that develop in time. The duration of exposure of the surface at various elevations of the pit as a result of the deepening of mining operations is not the same. On the sections corresponding to the upper part of the quarry, the time of the existence of rock outcrops is equal to the period of the complete development of open engineering, in the lower part—the time of working out the last horizon of the quarry. Undoubtedly, due to the fact that different parts of the bead serve different times with depth, the stability factor will also change.

It is known (Maystrovsky and Rozhkovsky 2010) that the mechanical strength of rocks depends on the speed of their loading. The higher the latter, the greater the strength of the rocks, so the well-known formula for determining the stability factor will change as

$$\eta(t) = \frac{\operatorname{tg}\phi \sum_{1}^{n} N_i \cdot + C(t)L}{\sum_{1}^{n} T_i}, \tag{5.7}$$

where

N_i—normal force on the ith platform,
T_i—The tangential force on the ith site;
n—the number of considered sites along the length of the sliding surface L.

It follows from expression (5.7) that the less the board is in time, the greater the cohesion of the rocks and the stability factor, i.e., at the beginning and by the end of quarrying the stability of the side will be different.

The use of equality (5.7) in determining the stability of the instrumental arrays is hampered by the difficulty and difficulty in finding the numerical value of the coupling as a function of the formation time of the rock outcrop. In addition, in the above formulas, the angle of internal friction is assumed to be unchanged, i.e., the limiting state of the rocks is described by a straight line, whereas in reality, the envelope of Mohr's circles has a curvilinear character.

It was proposed in Report No. 003219, KarPTI (2001) to use the mathematical expectation obtained from the distribution curves of the initial quantities when preparing the initial characteristics of the rock mass composing the side. At the same time, a rational level of risk is established based on the interaction of the assessment of the economic and psychological consequences of risk. To estimate the influence of time on the stability of career slopes, it was proposed in Nurpeisova et al. (2006) to use the laws of thermodynamics and to consider the bead as a thermodynamically isolated system in which the increment of the specific entropy of the side increases as it stands. At a critical level of entropy, the bead in the quarry can collapse.

A significant drawback in determining the stability of quarry slopes is that the strength properties of the rocks found in the calculations found in the laboratory under certain temporary conditions in most cases do not correspond to the time of the course of geomechanical processes in the instrumental massif, i.e., the conditions of similarity by the factor of time are not fulfilled. Previously received (Fisenko 1968)

$$\tau_i = \frac{1}{2}(\sigma_1 - \sigma_3)\,Sin\,2\alpha = C + \sigma_n\,tg\,\phi$$

$$= \frac{k \cdot T(\ln \dot{\varepsilon} - \ln \dot{\varepsilon}_0 + U_0/k \cdot T) \cdot Sin\,2\left(\frac{\pi}{4} - \frac{\phi}{2}\right)}{\gamma}, \tag{5.8}$$

where

- $\dot{\varepsilon}_0$—the maximum relative deformation velocity;
- U_0—initial activation energy of destruction; γ—coefficient of the structure of a solid;
- k—Boltzmann constant; T—absolute temperature of the test material,
- $\dot{\varepsilon}_0$—relative deformation velocity of rocks.

Analyzing formulas (3.2) and (3.3), we can conclude that the strength function depends on the time and temperature factors, namely the rate of deformation processes in the instrumental array, since the numerical value of the rock cohesion (C) depends on the relative strain rate and temperature of rocks.

The method for constructing strength passports (Beck 2005), taking into account the time factor, is based on applying the formula:

$$\frac{2\tau_k(t)}{\sigma_c(t)} = 1 + A\sigma_3^{0.67}, \tag{5.9}$$

where $A = k_n\,0.0893$ (k_n—a correction coefficient equal to that for rocks with c $\sigma_c < 30$ MПa – 0.6; $\sigma_c = 30$–80 MПa – 0.6–0.8; $\sigma_c = 80$–180 MПa – 1.0; $\sigma_c = 180$–200 MПa – 1.1; $\sigma_c = 200$–250 MПa – 1.2; $\sigma_c = 250$–300 MПa – 1.3; $\sigma_c = 300$–350 MПa – 1.4; $\sigma_c > 350$ MПa – 1.5). The uniaxial compression strength with time factor is Beck (2005)

$$\sigma_c(t) = \frac{k \cdot T(\ln\,t_0 - \ln\,t + U_0/k \cdot T)}{\gamma}. \tag{5.10}$$

The numerical values of the parameters entering into (5.10) were determined by the recommendations given in Okatov et al. (2003).

The diameter of the Mora circle in the stretch region can be defined as

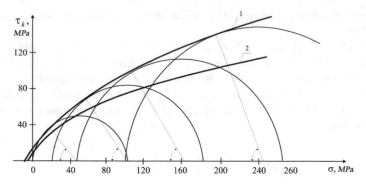

Fig. 5.12 Bends for the Koelgin marble at $\dot{\varepsilon} = 10^{-2}\ c^{-1} - 1$ и $\dot{\varepsilon} = 10^{-11}\ c^{-1} - 2$

$$\frac{\sigma_c(t)}{\sigma_p(t)} = 6.75 + 0.038 \cdot \sigma_c. \tag{5.11}$$

The envelopes constructed according to (5.8), as can be seen from Fig. 5.12 (Beck 2005), are of a curvilinear character, although in most cases in determining the stability of pit slopes the rock strength passport is represented as a straight line. It goes without saying that such an interpretation of the envelope to Mohr's circles can lead to significant errors in determining the basic mechanical characteristics, namely the cohesion and the angle of internal friction of the rock, which are necessary for finding the stability parameters of pit slopes. In connection with this important circumstance is the search for ways of finding these parameters and, ultimately, resistance to shearing with a wide variation of the principal stresses σ_1 and σ_3, taking into account the time factor and the curvilinearity of the bend.

From the expression it follows that $\tau_i = C$ at α and $\varphi = 45°$, then

$$C = \frac{k \cdot T(\ln t_0 - \ln t + U_0/k \cdot T)}{\gamma} \tag{5.12}$$

In formula (5.9), in determining C, the unknown quantity is the parameter γ, which is sensitive to the destruction method. To find it, we use the calculated method of constructing the rock hardness passports with allowance for the time factor proposed in Beck (2005), and we will determine from them the adhesion value C for different values of t for a rock with a uniaxial compressive strength at standard tests $\sigma_c = 44.8$ and 100.0 MPa.

The results of calculations (C) are summarized in Table 5.3 and are illustrated in Fig. 5.13.

According to Kasymkanova (2006), the numerical value of the coefficient of the rock structure for a given failure mode with the unchanged rock destruction mechanism (viscous or brittle) is independent of the time factor. Indeed, with a pure shift, the deviations of this parameter from the mean values of C = 8.03 MPa for

Table 5.3 Results of calculating the value of the adhesion versus time

Limit of rock strength on uniaxial compression, $\sigma_{cэc}$ (MPa)	The coefficient of the rock structure under exact shift, $\gamma * 10^4, \frac{joule}{mole} * \frac{mm}{kg}$,				
	$t = 1$ year	$t = 5$ years	$t = 10$ years	$t = 20$ years	$t = 30$ years
44.8	7.94	7.61	7.690	8.40	8.52
100.0	2.84	2.98	2.72	3.09	–

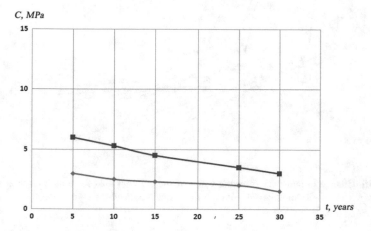

Fig. 5.13 Dependence of C, MPa on time t, years

rocks with $\sigma_c = 44.8$ MPa and C = 2.91 MPa for rocks with $\sigma_c = 100.0$ MPa constitute, in the first case, 5.5% other −7.0% (Fig. 5.14).

Analysis of Fig. 3.12 reveals a seemingly paradoxical fact: rock resistance to shear including a shear for strong rocks, is much lower than for weak ones. The possibility of the appearance of such a fact is clearly visible from the experimentally obtained plots $ln\ t\ (\sigma_c)$ (Korolev 1994).

Fig. 5.14 Dependence of the adhesion C (t) and the angle of internal friction ρ (t) for rocks 1 − $\sigma_{cж} = 44.8$ and 2 − $\sigma_{cж} = 100$ MPa

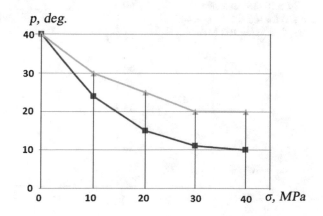

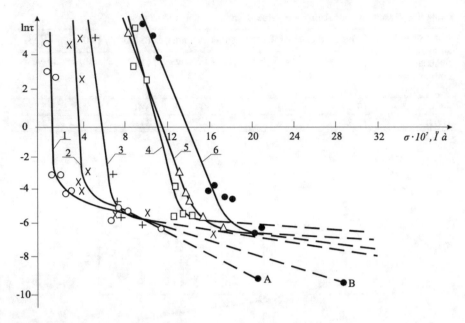

Fig. 5.15 Change in longevity from the value of the external load for limestone-shell rock—1, coal—2, siltstone—3, dolomite—4, Silurian limestone—5, sandstone—6

Figure 5.15 shows that when ln t increases, it is possible to intersect the ln t (σ_c) graphs with different strengths in standard tests. Calculations of σ_c by formula (5.12) do indeed show that when a certain time interval is reached, the strength σ_c (t) for a rock with a strength at standard tests $\sigma_c = 100.0$ MPa becomes smaller than for a rock with $\sigma_c = 44.8$ MPa (Fig. 5.16).

This phenomenon can be explained as follows. In the weak rocks under the influence of the load, initially sealing processes occur (slamming of microcracks,

Fig. 5.16 Dependence, σ_c (t) for rocks 1 − σ_c = 44.8 and 2 − σ_c = 100 MPa

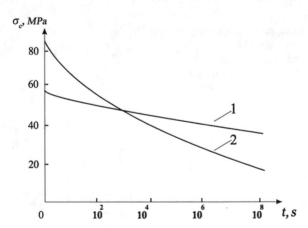

pores, etc.). Moreover, the longer the load acts, the stronger the test material becomes. In durable rocks, weakened processes (the formation of microcracks, their integration into larger cracks) are overcome.

In work (Korolev 1994), the possibility of taking into account the time factor in assessing the stability of rocks near both underground mine workings and instrumental massifs is shown. Its essence lies in the transition from stresses to the durability of rocks at t (the time before the beginning of the destruction of the rock) or to the relative rate of deformation of rocks at various points of the massif. The determination of the VAT of the array in the open-plan part of the device is also done by superposition and can be made from different marks: from the surface or from any point below the latter. This does not affect the distribution of VAT.

In this case, for the calculation of VAT as a starting point, an arbitrary mark of the side was adopted. The algorithm of the program is compiled in such a way that the calculations of the VAT of the array are determined within this mark, and then for the rest of the sideboard massif (Fig. 5.17).

Using the kinetic theory of the strength of solids, stress is transferred to the durability of rocks (the time of existence of the rock under load). The relationship between the strains acting in the array and the longevity of the rocks can be expressed as Galiev and Shamganova (2012).

$$t = t_0 \exp \frac{U_0 - \gamma\sigma}{kT} \tag{5.13}$$

where τ_0—period of vibration of an atom in a crystal grid, c; σ—arbitrary point stress of an array.

Taking into account that $\dot{\varepsilon} t = 10^{-2} = 10^{-2}$ [47]

you can go from the durability of the rocks to the relative speed.

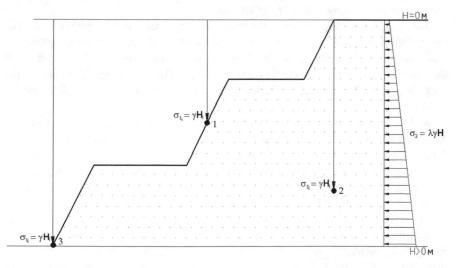

Fig. 5.17 The calculation scheme

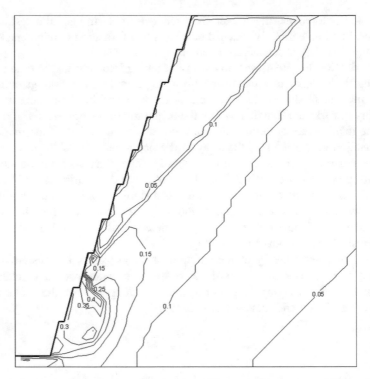

Fig. 5.18 Distribution of isolines of maximum tangential stresses in the quarry slope for $H = 300$ m (for heterogeneous environments)

Figures 5.18 and 5.19 show an example of assessing the stability of the side of the copper mine Corporation "Kazakhmys".

The weighted average value of the strength of the rocks of the massif on uniaxial compression is $\sigma = 120$ MPa. The strength of the array was corrected by the formula $\sigma_M = \lambda\sigma$ (where λ is the coefficient of structural attenuation). The stability of the pit was made from the depth of $H = 180$ m. The minimum durability of the rocks is $t_{min} = 10^{15}$ s, which indicates the stability of the pit at this depth (Korolev 1994).

At $H < 300$ m, no appreciable damage was observed in the instrumental massif. The most significant fractures in the lower part of the massif appeared at a depth of 300 m. Figure 5.19 shows the isolines of the maximum tangential stresses τ in MPa and in Fig. 5.20 of the longevity isolines t (in seconds) in the instrumental massif of the Spassky quarry at $H = 300$ m. The red line indicates the slip surface, determined by the method of Prof. G. L. Fisenko.

The advantage of this approach is that the slip surface is not predetermined, but is determined on the basis of the solution of the elastic problem of stress distribution in the instrumental array and the transition to durability or relative rate of rock deformation. Of course, the correctness of the assessment of the stability of career

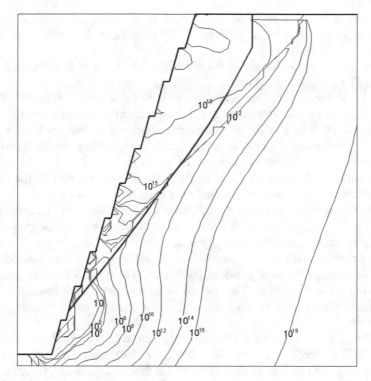

Fig. 5.19 Distribution of isolines of maximum tangential stresses in the quarry slope for $H = 300$ m (for a homogeneous environment)

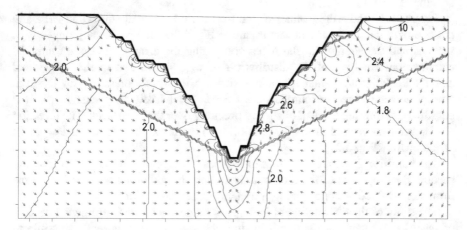

Fig. 5.20 Distribution of the maximum tangential stresses in the sideboard massif of the «Central» quarry

slopes will in many respects be determined by the accepted assumptions when solving an elastic problem and the reliability of entering data on the mechanical properties of the array.

For example, an assessment of the stability of the sides of the "Akzhal" quarry that is 300 m deep along the profile M. In the geological structure in the region of the profile M, the rock formations represented by sandstones, tuff sandstones with interlayers of tuffs and limestones are the main distribution. In some places, the bedrock has been interrupted by granite intrusion and, in the lower parts, over-lapped by Quaternary sediments—sandy loam, loams with gruss and gravel. The thickness of loose deposits varies from 0.1 to 2.0 m.

The Akzhal polymetallic deposit is located in the eastern part of the Akzhal–Aksaran crushing zone and is confined to a latitudinal-extended crushing zone in the carbonate-terrigenous deposits of Famen and the lower tour. The rocks composing the ore field are broken by a series of small intrusions, stocks, granite dikes, diorite-porphyries, and diabase porphyrites.

The main ore-bearing rocks are massive limestones, which are framed by the lower and upper horizons of siliceous limestones. These rocks form the composite part of the Akzhal anticline. The deposit is located within the linearly extended ore-bearing zone of the latitudinal strike, which is confined to the crushing zone in the anticline.

The ore zone with industrial mineralization can be traced at a distance of 5 km. By fall, it is explored to a depth of 550 m. The thickness of the ore zone varies widely reaching 70 m. The average thickness of the ore zone is 15–20 m. The fall is steep, the south one is up to the opposite. The ore zone of the deposit is conditionally divided into three sections—the Western, Central, and Eastern. The first section has been worked out, the Central and Eastern sections are currently being developed.

Figures 5.19 show the isolines of the maximum tangential stresses in MPa and the arrows of the displacement vector, and in Fig. 5.20—the isolines of longevity in seconds and the strength of the rocks composing the array of arrays on uniaxial compression. Analysis of the distribution of τ_{max} shows that in the ledges and directly in the instrumental array, the surfaces in which stresses with different numerical values are concentrated are traced. Most likely they characterize the possibility of forming a prism of sliding precisely over these surfaces. The direction of displacement vectors shows the possibility of extrusion of rocks in the lower part of the quarry (Fig. 5.21).

The transition from stress isolines to isolines of longevity allows one to take into account the properties of the rock massif and obtain additional information on the formation of fracture zones taking into account the time factor.

The study of the distribution of isolines of longevity in the instrumental massifs of the Akzhal quarry at the depth in question shows that it is really possible to distinguish the prism of collapse from them, within which deformation and disintegration processes take place most actively.

With increasing depth, the risk of collapse of rocks in the ledges increases. The formation of the sliding surface starts from the bottom-up. The time of its formation

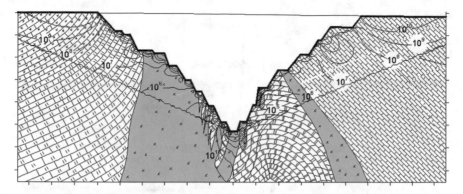

Fig. 5.21 Distribution of isolines of the longevity of the sideboard massif in the "Central" quarry

(in this case the stationary bead is considered) varies from 10^5 to 10^8 s (about 3 years).

The findings are characteristic only for the Akzhal quarry. The time of formation and configuration of the slip surface depend primarily on the natural (the strength of the rocks composing the array, their fracturing, the presence of tectonic disturbances, etc.) and mining factors (depth of work, slope of pit slopes, open-pit mining technology, and others).

The application of this technique will allow, at least in the first approximation, to assess the stability of the instrumental massifs, taking into account the depth of mining and the associated time that has elapsed since the beginning of the outcrop of the rock massif. A sub-array can contain geological discontinuous disturbances that alter its stress state, or a system of cracks that divide the rock mass into separate blocks, due to which the formation of sliding surfaces can partially or completely occur along the cracks present in the array. Undoubtedly, this technique works when the slip surface intersects with the cracks present in the array, the formation of slip micro-surfaces occurs due to breaking of molecular bonds or in the event that the cracks in the array are "healed", i.e., the material filling intercontact space, by strength is not inferior to the rocks composing the array, their influence can be neglected. If there is a continuous network of "untreated" cracks and their favorable orientation relative to the side of the quarry, the possibility of formation and sliding of the collapse prism will be determined by the processes taking place in the intercontact layer.

The size of the building blocks, in which the array behaves as a homogeneous one, can be determined from Fig. 5.22 (Kasymkanova and Kirgizbaeva 2004).

The formation of a prism slip does not yet mean the loss of stability of the array. In determining the safety factor, two distinct phenomena should be clearly distinguished: the formation of a collapse prism; slippage of the caving prism. In the first case, the deformation and disintegration processes are mainly due to the stressed

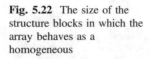

Fig. 5.22 The size of the structure blocks in which the array behaves as a homogeneous

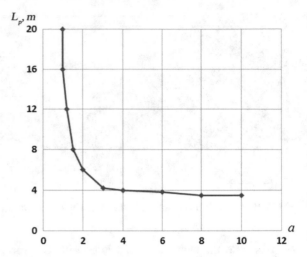

state of the instrumental massif and the properties of the rocks composing it, in the other—to slip with the destruction of the hooks of one surface relative to the other.

The resistance to the shear of one surface of cracks with respect to the other will be equal to $\tau_{ki} = \sigma_n \cdot f_{mp}$ (f_{mp}—the coefficient of friction between the macro-fracture surfaces). The study of the shear along the crack formed in the sample made it possible to isolate a quasi-stationary (creep) sliding with the destruction of the hooks and the stage of stationary slip of the banks with a smoothed surface at the outermost stage of deformation, the work of Mansurov. The results of researches testify to change of morphological signs, first of all a roughness of cracks, and consequently about inconstancy of factor of a friction f_{mp} in process of slide sliding slip. At the moment of crack formation, it has the maximum value, and by the end of the stationary slip the destruction of the protrusions is completed. The contacting surfaces of cracks are ground to each other and the strength properties of the array will be determined by the frictional force of the smoothed surfaces.

With a favorable orientation of the macro-fracture for the sliding of the formed collapse prism, the crack surface can undergo a change from rough to quasi The frictional forces that occur in the contacting layer can decrease. Laboratory tests have established that the strength characteristics of the rock contacts for rough and mirror crack surfaces may differ for diabase by 17%, and if there is a grind in the contact layer, the adhesion may be reduced by 33% compared to the rough surface and by 20% in comparison with the mirror contact.

In the presence of a weaker in the contact layer of the filler than that of the rock mass, the formation of the weakening planes will primarily occur along less stable elements, therefore, in calculating the stable parameters of the slopes and sides of quarries.

Thus, taking into account the stressed state of the rocks in the instrumental massif together with the time factor gives more objective information about the formation of a possible collapsed prism during various periods of quarry exploitation at the design stage. This approach allows us to simulate geomechanical processes and possible negative manifestations in the form of landslides at the design stage and determine the parameters of the stability of the pit slopes taking into account the service life.

References

Baikonurov OA, Melnikov VA (1970) Fundamentals of mining geophysics. Science, 326 p (In Russian)

Beck ASh (2005) The use of GIS technologies in the detection of active systems of cracks. In: Proceedings of International Scientific and Prak Conference "Agoshkov Readings, IPCON RAS, pp 91–95 (In Russian)

Fisenko G (1968) On the state and tasks of laboratory and full-scale tests of the strength and deformability of rocks, Tr. ATTENTION (In Russian)

Galiev SZh, Shamganova S (2012) Geomechanical justification of the boundaries and development of an automated system for corporate monitoring and management of the geotechnological complex of mining of gold-mining deposits Vasilkovskoe, Suzdal and others, Report on research. Works of Institute of Mining named after D. A. Kunaev, 67 p (In Russian)

Galustyan EL (1989) Artificial strengthening of unstable areas of quarry boards. Occup Saf Ind 12:23–25 (In Russian)

GOST 21153.0-75 (1975) Mountain rocks. Sampling and general requirements for physical testing methods. State Committee for Standards, 35p

Iofis MA, Grishin AV (2005) The nature and mechanism of the formation of concentrated deformations in the mold of shifting. Min Inf Anal Bull 7:82–86 (In Russian)

Jangulova GK, Tulebaev KK (2014) Conducting laboratory tests to determine the physico-mechanical characteristics of rock samples. Report on research. Funds IGD them. YES. Kunaev, 28 p (In Russian)

Kasymkanova HM (2007) A technique for studying the strength properties in a dump scientific and technical support of mining. Proc Min Inst 73:233–236 (In Russian)

Kasymkanova KhM, Kirgizbaeva G et al (2004) Methodological recommendations for the assessment of the stability of the massifs of the quarries Akzhal. KazNTU, 24p (In Russian)

Kasymkanova KhM (2006) Difficulty predicting large-scale destruction of the sides of quarries. Scientific and technical support of mining. Proc Min Inst 71:12–18 (In Russian)

Korolev YuK (1994) Opinion of a specialist. GIS-Review, pp 5–8 (In Russian)

Maistrovsky MA, Tulebaev KK (2010) Stability of the side of the quarry taking into account the internal dumping. Collection of works of Institute of Mining named after D.A. Kunayev 78:3 (In Russian)

Mashanov AZh, Nurpeisova MB (2000) Geomechanics. KazNTU, 124 p (In Russian)

Maystrovsky, Rozhkovsky GF (2010) Program Sustainability (Version 3.2—1974) (In Russian)

Methodical instructions (1988) Converters piezoelectric PRIZ-12. Chisinau, 123 p (In Russian)

Nurpeisova MB, Kasymkanova HM, Kirgizbaeva GM (2006) Methodical recommendations for assessing the stability of the angles of inclination of the ledges and sides of the quarry of the Zherek deposit, 24 p (In Russian)

Okatov RP, Nizametdinov FK, Tsai BN, Bondarenko TT (2003) Accounting for the time and temperature factors in the construction of rock strength criteria 2:38–42 (In Russian)

Popov II, Nizametdinov FK, Okatov RP (1997) Natural and technogenic foundations of management of stability of ledges and sides of quarry. Gylym 215 p (In Russian)

Report № 003219. (2001) Instrumental control over the stability condition of the sideboard massifs of the Sayaksky quarry. KarPTI, 25 p (In Russian)

Report No. 619-90 / 24. UNIPROMED. (1980) Investigate the stability of the sides of the ledges of the Kounrad pit and develop recommendations for the safe management of mining operations. Sverdlovsk, 180 p (In Russian)

Savich AI et al. (1969) Seismoacoustic methods for studying rock massifs. Nedra, 240 p (In Russian)

Tsai BN, Bondarenko TT, Malakhov AA, Amanbaev BZh (2005) Accounting for the rheological properties of rocks in assessing the stability of mine workings. Actual Problems of the Mining and Metallurgical Complex of Kazakhstan, pp 121–123 (In Russian)

Yakovlev VL (2009) State, problems and ways to improve open mining developments. Min J 11:11–14 (In Russian)

Chapter 6
Development of Solutions for: Hardening, Strengthening the Sides of the Quarry; Suppression Dust Formation on the Quarry Roads

6.1 Analysis of Existing Studies of World Experience in Developing the Composition of Solutions Used to Strengthen the Sides of the Quarry

With the increase in production and the depth of quarries, the urgency of the problem of increasing the stability of slopes increases. Research in this area includes the development of strengthening slopes and hardening of rocks (Kurmankojayev 2003). If, for the prevention of collapses and landslides, the artificial increase in the strength of large rock massifs is not yet practical, then the prevention of deformations of individual ledges and prevention of scree formation from the surface of slopes by artificial reinforcement is now being used in domestic and foreign quarries.

In most quarries, rocks forming nonworking beads are heterogeneous in structure and physical–mechanical properties. Even in relatively uniform eruptions of the rocks of the open field, there are usually 5–6 or more zones of tectonic disturbance. Depending on their size and spatial orientation, they have a more or less significant effect on the stability of the beads and ledges.

With the help of strengthening, it is possible to slow down the processes of weathering and shedding of rocks to prevent the collapse of the ledges and the scattering of rocks from the surface of the slopes (mechanical strengthening and arrangement of protective coverings of slopes).

The use of artificial strengthening of rock massifs allows increasing the angles of slopes on sites with less stable rocks. The feasibility of strengthening is established by technical and economic calculations. The beginning of systematic research in this field was made in the laboratory for the stability of the sides of the VNIMI quarries, where the practical experience in strengthening quarry slopes was analyzed, theoretical and laboratory studies were carried out, on the basis of which were published "Methodological guidelines for the artificial strengthening of slopes of rock and semiccale rocks in quarries."

K.-K. Kassymkanova et al., *Geomechanical Processes and Their Assessment in the Rock Massifs in Central Kazakhstan*, SpringerBriefs in Earth Sciences, https://doi.org/10.1007/978-3-030-33993-7_6

The ways to strengthen career slopes are divided into four main groups (Eliseeva and Rukavishnikov 1977):

– mechanical methods of strengthening;
– methods aimed at improving the physical and mechanical properties of rocks;
– methods that ensure the isolation of the surface slopes from the effects of unfavorable external factors;
– mixed methods.

Strengthening unstable slopes by mechanical methods is based on the redistribution of stresses in the rock massif.

The pressure of the collapse prism, the reinforcement structures transmit a stable part of the massif. A necessary condition for the use of mechanical methods is the existence of a solid, stable massif that is outside the zone of possible collapse.

As a means of mechanically strengthening slopes, structures (piles, veneers, rods, cable ropes) and solid structures (retaining and protective walls, buttresses) are used. The former are used primarily for strengthening slopes of rock and semi-rock rocks, the latter for preventing landslides and strengthening clay rocks and filter slopes.

The second group of methods is aimed at improving the physical–mechanical properties of rocks by strengthening them with the help of various binders and cementing materials injected into the array to restore broken structural bonds in fractured rocks or to create new additional bonds in loose rocks.

Coating of the surface of slopes with insulating materials is carried out under the condition of intensive weathering of rocks in slopes after their exposure. Coverings are intended for the isolation of an array of rocks from the influence of an environment.

In complex engineering–geological conditions, the complex is strengthened in a variety of ways.

For example, the method of cementation of fractured rocks is used in combination with mechanical methods of strengthening piles or bars (Turbovich et al. 1971).

A. The mechanical method of strengthening is used in arrays:

(1) With an underdeveloped fracture, intersected by unfavorably oriented weakening surfaces, falling toward the worked out space at an angle of 200–500.
(2) Large-block, slightly arched massifs, slate-layered rocks falling toward the worked out space at an angle of 400–600. They are strengthened with bars and flexible cable ropes.
(3) Strongly fractured weathered rocky and semi-local rocks strengthened by protective walls.
(4) Disrupted arrays of a complex structure with interbedded rocks, clayey and friable moistened rocks. Ways to strengthen the retaining walls and buttresses.

B. Strengthening of the massif (improvement of physical and mechanical properties of rocks):

(1) Rock and semi-broken disturbed and fractured massifs, coarse-grained sands without clay, injection into an array of reinforcing solutions of polymeric materials (resins), cementation.
(2) The rocks are sandy and loess (the filtration coefficient is 0.1–5 m/day).
(3) The rocks are water-saturated clay (the filtration coefficient is less than 0.01 m/day). Strengthening by electrochemical treatment (electroosmosis).
(4) The rocks are loess-like, loamy, and clayey with an air permeability coefficient of 0.2–0.4 cm/s. A heat treatment method.
(5) The rocks are super sandy, loess-like, clayey with a coefficient of porosity of more than 0.1, compacted by the energy of the explosion.

C. Insulating and protective coatings:

(1) Strongly fractured rocks are prone to intensive weathering or leaching. Mounting spatter-concrete on a metal grid.
(2) Strongly fractured rocks, sandy and graveled slopes are reinforced with resins.

D. Combined strengthening:

(1) Complex engineering and geological conditions. Fastening mechanical strengthening with hardening, mechanical strengthening with insulation.

Hardening of the rock massif on the weakened sections is achieved by introducing into the cracks of the massif of substances, which after hardening and setting the rock significantly increase its resistance to shear. The introduction of reinforcing material into the massif is carried out under pressure, and cementitious solutions, silicates, and polymer resins are used as the hardening material (Bek et al. 2006).

The most widespread among hardening methods was rock cementation. Cementation is carried out when carrying out excavations in aquifers, strengthening unstable and disturbed massifs. The field of its application is an array composed of rocks from highly fractured rock and pebbles in the presence of cracks in the width of not less than 0.15–0.20 mm, providing access to the cement slurry in the fracture. Species should have a specific water absorption of at least 0.01 L/min.

The best cementation effect is achieved in fractured rocks (sandstones, argillaceous and sandy shales, limestones, granites, granite-gneisses, gabbro, argillites, siltstones, etc.) in the absence of clay filling in cracks. Cementation is used not only to create waterproofing curtains in the foundations of dams but also as a method of strengthening the rock massif.

Through cementation, a strong mating of the body of the dam with the bank of the river is achieved and the formation of a reliable stop before filling the upper tail.

Cement solutions are injected into the array through the wells and should have pour points that ensure its penetration through the cracks to the design distance. The setting time of the cement with the rock is regulated by adding to the solution soda, calcium chloride, and other substances.

One of the conditions preventing cementation is a weak resistance to the action of aggressive waters. Some groundwater contains soluble mineral salts and acids that cause corrosion of cement and concrete.

Glynization is used to fill voids in karst rocks and in rocks with large fractures. The drawbacks of this method include a large consumption of oil-well material, a considerable duration of the process, a weak resistance of the clay slump to external pressure, the inadvisability of plugging fine cracks. Advantages of the method are: the possibility of using local plugging material (clay), the ability to withstand the action of corrosive waters, eroding even special concretes.

To strengthen the rock massifs, a method of silicization is also used—the introduction of sodium silicate and calcium chloride into the array.

Since the permeability of rocks is too small, the introduction of silicates into the array can be carried out by an electrochemical method. In this case, the introduction of solutions is not made by injecting them into the wells, but by creating a difference in electrical potentials between adjacent wells.

Bituminization—the filling of voids and cracks by injecting molten bitumen into them. In the liquid state, the bitumen penetrates into the cracks with an opening of more than 0.2 mm. However, even in a solid state, bitumen retains ductility, and at a certain pressure begins to flow and separate from the cracks.

Artificial freezing of rocks is used when drilling shafts and other mine workings in aquifers. The technological process is very expensive and takes a long time.

To strengthen the weak watered sandy-argillaceous rocks, the electrophysical method of electroosmosis is used, i.e., impact on the breed of direct current, resulting in the drying of the array, which increases its strength characteristics.

In recent years, intensive research has been carried out on the artificial strengthening of fractured rocks using synthetic resins.

Polymer resins play the role of synthetic glue, fastening blocks of rocks, broken by cracks. Reinforcing compounds are complex mixtures including synthetic resin, hardener, setting accelerators, stabilizers, solvents, and other special additives.

For the preparation of solutions, various resins are used—phenol aldehyde, amino aldehyde, polyester, epoxy, silicone, polyurethane mixtures, etc.

The technology of rock hardening by injecting fastening compounds has been tested in the mines of Donbass, Kuzbass, at the mines of PO "Apatite", Krivbassrud, Ural, Norilsk, Kola, and other polymetallic deposits.

Injection hardening technology is widely used at mining enterprises in Germany, the USA, Great Britain, France, Japan, the Czech Republic, Belgium, the mines of Canada, India, the USA, South Africa, and other countries.

A lot of experience in the use of polymer resins to strengthen weak areas in the rock mass is available, in particular, in Germany. Polymer resins are used here to eliminate convergence, strengthen the soil in loose rocks, outstrip fixing of mine workings, fill voids (cupolas) in tunnel construction, in mines and mines.

The most promising material for hardening the rock massif are polyurethane resins. They are the material formed as a result of the reaction of polyisocyanates with polyol resins. Prior to mixing, the components are fluids with a relatively low viscosity; when combined, the components retain their flow properties for some time, then the composition hardens with increasing volume. The main advantages of polyurethanes are high strength, low molecular dispersion, allowing the composition to penetrate into cracks with a width of opening less than 0.01 mm, high adhesion bond strength ("adhesion" from the Latin word adhaesio—adherence) of the boundary layer "polymer—rock", high coefficient of foaming (from 1.5 to 8), low toxicity, acceptable (from 3 to 45 min) gel time. Possessing high penetrating ability, the hardening composition, when injected under high pressure, fills 90–95% of all cracks in the array. The hardened polyurethane has residual plasticity, which makes it possible for the hardened mass to deform without fracture and withstand seismic loads. Foaming polyurethane creates an additional spacing effect, which increases the bond between the blocks of the array and improves its strength properties.

Distinctive features of polyurethane, which determine its advantages over other types of resins, are high adhesion with rocks under pressure curing, adjustable pour point and strength, exothermic nature of the polymerization reaction of the components (self-heating), and residual plasticity, which allows to not lose the connections of the hardened rocks at explosive works and processes of rock shifting.

The above properties of polymeric polyurethane resins indicate the expediency of using them to strengthen the array of rocky and semicatal fractured quarry rocks.

The modern approach to the solution of the problem of maintaining the instrumental massifs is based on the maximum use of the intrinsic carrying capacity of the surrounding massif. One of the effective ways to increase the load-bearing capacity of a rock massif is, as is known, to strengthen the bonds along fissured contacts of structural blocks in the depth of the massif and its part by injecting synthetic resins or other types of binders, introducing reinforcing elements or in a combined way. This makes it possible to effectively solve the main tasks of developing mineral deposits—improving the extraction of minerals, reducing losses, and improving the safety of mining.

It should be noted that there is no significant experience on hardening of the massif by resin injections in Kazakhstan and CIS countries. This problem requires additional theoretical and industrial–experimental studies.

To strengthen the array of strong fractured rocks, which represent the majority of ore deposits, it is effective to use synthetic resins that have sufficient strength, low (molecular) dispersion, allowing the resin-based composition to penetrate into thin pores, and cracks with an opening width of less than 0.01 mm.

The technology of rock hardening with synthetic (urea) resins was first applied by the MHTI named after DI. Mendeleev and VNII oil and gas for the isolation of reservoir waters by creating watertight screens in oil wells. Since 1956, this technology is engaged in the Institute of Mining named after A.A. Skochinsky (Nizametdinov et. al. 2007).

The technology of rock hardening by injecting fastening compounds based on synthetic resins is based on the creation of two-, three-, and more-component solutions of chemical compositions, the combination and mixing of which leads to the formation of a solid with high adhesion to rocks and other physicochemical and mechanical properties corresponding to requirements of mining. The curing of the components to be joined occurs as a result of polymerization or polycondensation reactions, characteristic for the curing of synthetic resins with the introduction of hardeners, hardening of the mineral solution with water or other special solutions (Ozhigin 2009).

The experience of many industrialized countries shows that instead of erecting traditional types of support in the underground mining of deposits, in many cases it is advisable to strengthen the weakened fracturing of the contour array by polymeric compositions that make it possible to form a monolithic shell with a high bearing capacity on the contour of the massif.

In resin injection practice, one of the important criteria for selecting and evaluating the method for hardening rocks is the size of the cracks and the comparability of these dimensions with the size of the particles of the hardening solution. The finest cracks occurring in rocks have an opening amount of about 0.01 mm. For the successful use of hardening compositions, it is necessary that the particle sizes of the reinforcing agents are 2–3 times smaller than the minimum crack opening.

Achievement of such small sizes of strengthening particles is possible when using solutions of chemical substances in which the substance is in a molecular state and the particle sizes are commensurable with the dimensions of the molecules themselves. Consequently, this condition is met only by chemical methods of hardening.

Reinforcing compounds are complex mixtures, including synthetic resin, hardeners, setting accelerators, stabilizers, solvents, and other special additives.

Compared with cement slurries, polymeric compositions have a higher penetrating power, and the duration of their hardening and other physicochemical and mechanical characteristics can be varied within wide limits.

The reinforcing composition is introduced under pressure into the fractured massif remains for a while liquid, then (after a period of gelation) it gradually loses its fluidity, becomes viscous and, after a short time, rigid, insoluble, and practically non-melting. About 70% of the strength of adhesion to the array, this material accumulates after several hours of injecting solutions. At the same time, the supply of liquid components of the fastening compound to the fractured massif is carried out with the help of special technological equipment including pumping pressures, main high-pressure hoses, connecting and shutoff-mixing valves.

Various thermosetting resins are used for the preparation of solutions—phenol–aldehyde, amino aldehyde, unsaturated polyester, epoxy, silicone, polyurethane mixtures, and others.

Most often, polyurethanes are used, which can repeatedly increase their volume, quickly harden, have plasticity, and high adhesiveness. Due to the increase in volume during foaming, polyurethane effectively fills the cavities and cracks in the rock massif, but its cost remains high.

The preliminary stage of hardening includes the assessment of joint of rocks, compilation of mineralogical characteristics, the designation of the form and composition of the formulation, the choice of technological equipment, the setting of injection parameters (pressure and injection rate are determined by experimental injections).

The radius of propagation of the fastener composition along the cracks of the array is determined experimentally every time. On the basis of the obtained data, the parameters of injection holes or wells are set at this particular hardening site.

In the practice of resin injection hardening of rocks, three schemes of injection of the array are distinguished: preliminary, advanced mining work, simultaneous, and subsequent.

Preliminarily, the rocks are strengthened before mining operations which is carried out in them or before the commencement of mining operations, which should provide favorable and safe conditions for mining operations in advance. Simultaneous strengthening is carried out in the process of excavation, and it is part of the tunnel cycle.

The subsequent strengthening is used to strengthen the rocks after the works have already been completed.

According to the method of supplying the ingredients and the place of production of strengthening solutions, three technological schemes are also distinguished: one-solution, multi-solution, and mixed.

In the first case, the solution is prepared in advance before it is fed into the array, using for this purpose a special container into which the individual components are fed and then mixed. According to the second scheme, the ingredients of the mixture are fed separately, mixing them in an injection well (drill hole). The third scheme provides for the supply of the ingredients of the solution immediately before injection into a special mixer, from where the mixture is pumped into the array.

The first scheme is the simplest, but with its use, it is necessary to choose a component ratio at which the duration of the gelling of the mixture would be at least 30–40 min. Otherwise, a portion of the precooked portion can harden before the end of the injection, which will lead to clogging of the equipment used and its failure.

With the simultaneous injection of two solutions, it is difficult to ensure a thorough mixing of the components inside the injection hole, which worsens the required quality of the composition.

In practice, a mixed injection scheme is most often used. It allows you to easily change the duration of the gelling process and the technical properties of the hardening mixture, while (in comparison with the single-solution scheme) the bond strength of the rocks is increased. With a separate injection of tar and hardener, less powerful pumps can be used. Injection compositions in the array are usually injected with pumps developing pressure from 3 to 30 MPa.

In recent years, no-pump injection schemes have been developed. In the injection well, special ampoules are placed with the ingredients of the strengthening compound, which penetrates into the array by exploding the explosive charge placed in the same well, or by multiplying the volume of the mixture in a closed volume of the well after mechanical destruction of the ampoules and mixing of the components of the composition.

From the point of view of the technical level and development trends of the issue under consideration, the foreign experience of injection reinforcement of rocks with synthetic resins is of great interest nowadays. The most widespread strengthening of fractured massifs by injection of synthetic resin compositions is used in the development of coal deposits, but in recent years, resin injection methods have also been used in the development of ferrous and nonferrous metals.

The effectiveness of the use of polymer compositions to improve the stability of fractured rocks is determined, first of all, by the strength of the adhesive bond between the boundary layer of polymer and rock. To the greatest extent, this requirement is met by epoxy resins. Currently, given the urgent need for epoxy resins in the mining industry, the USA has established more than 130 modifications. Resins on an epoxy basis are used to harden the roofing rocks of excavations in coal mines.

In the CIS countries, several grades of epoxy resins have been developed and are being produced for the reinforcement of the roof of the cleaning chambers and preparatory workings in the mines with steel–polymer and polymer anchors. Their use in underground conditions is limited due to high viscosity, high cost, and toxicity (Nizametdinov et al. 2013).

The scientific and experimental research carried out by the Moscow State Mining University allowed us to establish the field of rational application of the resin injection hardening method and on this basis to systematize possible hardening objects in the development of ore deposits.

The use of technology for hardening rocks at these sites by injecting bonding compounds in the developed versions can yield high results.

According to the studies carried out (Nizametdinov 2012), injection hardening is expedient in rocks that have their own strength at least at the molecular level. These include all strong fractured, medium-strength rock formations including shales, sandstones, and the like. Resins can be strengthened fractured massifs with medium block and small block structure, weak mutual engagement of separate parts up to 0.5 m, with fracture intensity from 0.1 mm/m and more, crack opening width from 0.01 mm, specific water absorption from 0, 0.1 L/min and above, rheometric permeability of rocks from 0.15 MPa/(m min) and higher.

Some results on resin-hardening of finely fractured massifs composed of alkaline rocks are described in the Report on the topic, №1.758.03 (2003).

In general, it can be noted that polymeric resins having a homogeneous structure and high penetrating power provide the best reinforcing effect, better adhesion to the rocks during curing. Therefore, they can be considered as the most promising in solving the problems of increasing the stability of rocks by the method of injection hardening. Of all the polymer resins, polyurethane and urea resins are the most effective for this purpose.

Smoin injection hardening is at the stage of active development and, first, the way and methods are very diverse, and second, there is a noticeable tendency to develop more and more reinforcing substances with improved physical and mechanical characteristics and other chemical and technical indicators.

Strengthening the rock massif of synthetic rocks with synthetic resins is a new trend in the theory and practice of managing the stability of quarry slopes.

The idea of strengthening resins is associated with scientific and technological progress in the field of chemical technologies in the second half of the twentieth century. This allowed, on the one hand, to obtain polymers with predetermined properties and, on the other hand, the production of resins on an industrial scale led to a decrease in their cost, which made possible the wide use of synthetic materials in the mining industry to solve the problems of strengthening rock massifs.

6.2 Development of Solution Compositions for Hardening, Strengthening of Quarries

One of the known ways to strengthen fractured rocks is their carburization. In quarries rock carburizing is applied as follows. From the upper landing site, vertical and inclined wells are drilled at a distance of 4–6 m from each other and the cement mix is injected into them until the array is completely saturated. Cement mortar is prepared on the basis of water and cement. The following solutions are known for strengthening fractured rocks:

(1) containing cement, water, and calcium chloride in an amount of 1–2.2% of the mass of cement (Nurpeisova et al. 2002);
(2) solution invented by the Donetsk Coal Research Institute, containing phosphogypsum binder, fast-hardening urea–formaldehyde resin KFB, oxalic acid, water (Rudnensky 2003).

These solutions have a high cost, and the second solution must still be prepared directly from the well.

As the closest analogue, a solution was taken, which includes:

1. Portland cement M 400 Karaganda cement plant with the characteristics: setting time—the beginning 2 h 50 min; the end: 3 h 40 min; normal density—25.5%, water–cement ratio <0.4.
2. Tails—waste from the enrichment plant of the Balkhash Mining and Metallurgical Combine (BGMK).
3. Calcium chloride ($CaCl_2$), accelerator.
4. Dispersive polymer powder AM 2572—products of "Clariant" (Germany).

Dispersion polymer powder tylos MB 15009—products of "Clariant" (Germany).

The main task is to create a mortar for strengthening rocks that has low cost, sufficient fluidity to fill small cracks, and adhesion to rocks, high strength. To achieve the goal, a solution is proposed for strengthening fractured rocks containing cement, filler, and water. As a filler, it is proposed to use tailings of concentrating factories, which are a large-tonnage production waste, and to allocate large areas for their storage to reduce the cost of the solution. Keeping tails causes great harm to the environment. The negative effect of tailings on the environment is manifested in the pollution of atmospheric air, underground and surface waters, soil, and vegetation cover with harmful substances.

And additionally, a dry superplasticizing additive NEOLIT 400, produced by Neochim (RK, RF) with high water reducing capacity and makes it possible to reduce the water-binding ratio in systems by more than 20%. With a decrease in the water-binding ratio, the durability and density of the solution being developed are increased, with a simultaneous reduction in shrinkage and creep strains when the strength of the solutions is set.

The additive is well compatible with Portland cement.

Concerning the components, the following is the wt%: cement up to 37%, tailings concentrators up to 52%, Superplasticizer NEOLIT 400 0.11–0.16, and the rest water.

This ratio of components was obtained experimentally and is optimal. On the one hand, the necessary fluidity of the solution and adhesion, and on the other hand, to obtain a material of the required strength after setting it with rocks.

Use of tailings of concentrating factories allows to lower the cost price of a structure and to raise durability of material. Increasing the amount of cement is more than 35%, and the superplasticizing additive NEOLIT 400 more than 1% increases the cost of the composition. The reduction in the amount of cement is less than 30%, and the superplasticizing additive NELOIT 400 of less than 0.9% leads to a decrease in adhesion of the composition with rocks and a decrease in the strength of the resulting material. Increasing the number of tailings of concentrating mills more than 55% reduces the fluidity of the solution and its adhesion to rocks, and a reduction of less than 45% increases the cost of the composition.

To determine the strength of the mixture, 4 × 4 × 16 cm samples were molded and compacted on the vibrating pad for 45 s. After 24 h, samples are taken from the molds and stored under wet conditions for 28 days (the starting value), and then physico-mechanical tests are performed, the results of which are presented in Table 6.1.

Table 6.1 Physical and mechanical properties of the solution

	Composition of solution, wt%				Indicators		
	Cement	Tails rock massif	NEOLIT 400	Water	Compressive strength, MPa	Bending strength, MPa	Sediment of the cone, mm
Sample 1	32	52	0.16	15.9	32.4	4.3	150
Sample 2	33.4	49.3	0.13	16.3	35.7	5.1	146
Sample 3	37	47	0.11	16.9	36.9	5.7	142

The data of the studies confirm that the proposed composition of the solution for strengthening the fractured mountain massif should be in the following ratio: mass%: cement 32–37, tailings of concentrating plants 47–52, superplasticizer NEOLIT 400—0.11–0.16, the rest water.

All components are loaded into a concrete mixer and mixed thoroughly with the addition of water.

Thus, the use of the solution described above provides strengthening of the weak sections of the quarry sides and allows to significantly reduce the harmful effect of waste from concentrating factories on the environment.

6.3   Analysis of Existing Studies of World Experience in the Development of Solutions for the Suppression of Dust Formation on the Roads of the Quarry

Open mining takes the leading place (more than 70%) in mining. Their significant shortcomings are significant violations and contamination of the work area by dust emissions. When developing deposits, a significant number of different sources of dust emission (drilling and blasting, excavation, transportation of rock mass, etc.) function. Dustiness in the conduct of work in massif can range from 0.5 to 10,000 mg/m. Dust loading leads to increased morbidity, mortality, and significant disability. The main diseases are special forms of lung diseases, and the main risk group is employees of enterprises with 15 years of experience in hazardous work (Rudnensky 2003).

Despite the considerable scale of the conducted studies and the proposed constructive solutions, the practical results are modest, especially for enterprises operating for most of the year in conditions of negative air temperature. This is due primarily to the inability to use traditional dust suppression in these conditions. In this regard, the search for rational means and ways to reduce dust emissions into the atmosphere is still an urgent task, especially for quarries.

The largest contribution to air pollution by dust emissions during the development of minerals is made by unorganized open sources of dust emission, the main ones being the dusting surfaces of roads, tailing dumps, ore and rock dumps, etc. (Pevzner 1992).

Analysis of existing studies shows that the suppression of dust on gravel and subterranean technological roads in quarries, then the most effective method of control is the spraying of the liquid dust suppressant. Even after a single application of the substance, the amount of dust rising above the road after the passage of the vehicle is significantly reduced, the need to use the watering machines all day long disappears. In the first year, the costs of dust suppression measures using chemical composition usually do not exceed the costs of traditional dust suppression by water. In the second and subsequent years, "supportive" treatments are carried out, gradually more and more rarely. Since the second year, these costs have been significantly reduced, and in a few years, the total savings can be up to 75% compared to dust suppression by water. Upon repeated application, the substance is combined with the previously introduced, thus gradually accumulating, dust particles that gets enveloped and become heavier, much less small particles remain that can fly into the air in the form of dust. This trend will continue, and the budget for dust suppression will be only a small fraction of what was spent for daily irrigation with watering machines. Some chemical compositions show excellent results even with very heavy cargo traffic along the treated quarry road (Zaslavsky et al. 1990, No. 1627714).

Binding chemicals are a universal remedy. They can suppress dust during construction, protect the soil from erosion. Some chemical reagents serve not only for dust suppression, but also reduce coal losses due to weathering during transportation by rail, and also prevent the risk of spontaneous ignition of coal in storage facilities and the explosion of coal dust. Stacked bulk materials can be treated with binders. Substances that form the crust are especially advantageous for processing stacks that are not in operation for a long time—the crust requires no maintenance and the stack does not need to be watered. In order to protect the cement–polymer and concrete floors from the formation of dust, various impregnating compounds are used.

One of the most effective and applicable worldwide methods of dust suppression is the spraying of water-based polymers, a vinyl–acrylic emulsion. Most of the polymer products used to stabilize and strengthen the soils are copolymers based on vinyl acetate or acrylic.

It is recommended that you first align the road with a grader, since most acrylic copolymers work best on the newly grounded soil surface. After the emulsion is applied, the water evaporates and the product hardens. Typically, the polymer is sprayed at least twice. The resulting reinforced layer is highly frost-resistant (remains elastic and does not crack at temperatures down to −30 °C) and is resistant to aggressive media, high elasticity, reliably protects the soil against wind and rain erosion.

Copolymers can be produced in the form of a powder. Such substances are either scattered on the surface being treated, or mixed with the ground, or dissolved first in the liquid.

Depending on the amount of substance introduced, the time of protective action is regulated—from weeks to several years. There are chemical agents (as claimed by their manufacturers), which after a single application (treatment of the surface of the stack, tailings) perform the function of dust suppression for 8–10 years.

If a technology is used in which the soil is loosened, mixed with the emulsion and compacted by the roller, also applying a thick film on the surface, the soil can be given the required strength, up to a value comparable to the strength of cement, the hardened layer will last for several years, serving as a dust suppression means for the road, only once a year will need to further spray the product over the surface. When the potholes begin to form on the road thus treated, it is sufficient to simply cover them with a mixture of soil and emulsion, the mass will adhere to the surrounding soil and become part of a flat road.

Products from natural organic polymers that are applied to the road surface are also used. Such polymers are nontoxic and completely biodegradable. This means that after the polymer expires, the soil returns to its original state.

To enhance the effectiveness of water in dust suppression, polymeric astringent and chemical wetting surfactants such as soaps and detergents can be added that reduce the surface tension of water and provide moisture to more soil particles of available moisture, as well as organic humidifiers. An ionic reagent can be added to the water, which will increase the ability of the drops to combine with dust particles. As manufacturers of such chemicals say, sometimes 2 mg of substance per 1 м3 of water is enough to reduce the amount of dust in the air by half. The effectiveness of the action of water increases to 50%, the water evaporates more slowly, the surface (roads or areas) remains moist longer.

Another substance is magnesium chloride, (bischofite, $MgCl_2$)—an effective solution for dust suppression. Ground and gravel technological roads after grading can be treated with magnesium chloride (and other hygroscopic salts, such as calcium chloride, hydrated lime (calcium hydroxide), and sodium silicates (liquid glass), which absorbs moisture from the air and moistens the surface of the road. Materials designed to combat dust, bischofite absorbs moisture from the air even in hot weather, which is enough to maintain the optimum level of moisture in the particles of the road. The layer treated with bischofite is further compacted (Kasymkanova et al. 2010).

Bischofite binds dust particles, which are formed due to the wear of the road surface, and does not allow them to rise into the air. Experts argue that depending on the intensity of traffic and the weather conditions of single use of bischofite is enough to dedust the road for a period of 3–6 weeks to several months, sometimes it is enough to apply once a year, and to suppress the dust of open storage of any dusty material—up to 18 months. The service life of roads increases by 2–3 times. Economic analysis shows that the use of magnesium chloride for surface treatment of roads reduces the annual cost of their maintenance by 30–40%. This substance has very little impact on the environment.

Bitumen–water emulsion with many additives is actively used for dust suppression. Bitumen resins collect individual dust particles in agglomerates rather than from a film like acrylic copolymers, so resins can suppress the formation of dust much longer than a copolymer film that quickly breaks down. As a result of using this composition, the use of water and watering machines is reduced by 80%: the composition should be applied twice a day with dilution in the water at a ratio of 1:40. The composition also increases the strength of the road, resulting in less formed ruts, due to which by two-thirds the costs for its repair and losses from the temporary closure of the road are reduced (Demin and Aleksandrov 1983). However, some experts argue that bitumen emulsions are too expensive because they require constant care when operating the road in quarries.

However, experience is the criterion of truth. In the United States, studies were conducted on the effectiveness of the dust suppression composition based on bituminous resin. The track was treated with a dirt road, and then for about 4 months, about 7,000 cars passed through it. After that, the content of dust particles of 10 and 2.5 microns (PM10 and PM2.5) was compared in the air over the treated and untreated roads. Above the processed road, the content in the air of PM10 was 86–98% and PM2.5 was 83–97% less than over the area not treated (Kasymkanova and Jangulova 2016).

Some experts argue that all chemical dust suppression products do not allow obtaining any lasting effect, require additional treatments, are not harmless to the environment, sometimes require heating (i.e., additional costs) in the distribution, easily washed out by atmospheric precipitation, weathering and forfeits to mechanical influences (Bekbergenov et al. 2016).

6.4   Development of the Composition of Solutions for Dust Suppression on the Quarry Roads

For the purpose of dust suppression on the quarry roads, as well as to prevent scree from the pit slopes placed in the limit position, the extent to which patents for the development of a solution based on the composition for fixing the dusting surfaces of dumps and other objects, as described in, is considered below (Kassymkanova et al. 2016).

The composition contains an adipic alkaline plasticizer (a waste of caprolactam production), Klevanoysky powdered clay extracted in passing with iron ore, and water at the following ratio of these components %: plasticizer adipic alkaline 10–15; Kvelavian powdered clay 15–20; water the rest. A coating obtained from a known composition has insufficient strength and durability.

In the composition for fixing the dusting surfaces of tailing dumps and other objects, given in the patent, contain, %: polyvinyl butyral 50–70; tails of enrichment the rest (Kasymkanova and Jangulova 2016).

The mixture is spread evenly on the area to be coated and with the help of gas burners, it is heated to the melting temperature of polyvinyl butyral 250–300^0 C. The strength of the coating is 10.3 MPa. The disadvantage of this composition is the insufficient strength and durability of the coating obtained, as well as the laboriousness of its application.

The composition for fixing the dusting surfaces of tailing dumps and other objects, described in the description (Kasymkanova 2016), contains clay, tailings, straw or cane and water at the following ratio %: clay 70–78; tails of enrichment 2–7; straw or bulrush 1.6–2; the rest is water. From this composition, pellets are made and placed on the dusting surface with a spreader of organic fertilizers. A disadvantage of the known composition is the insufficient strength and weather resistance of the coating obtained, as well as the laboriousness of its application.

When developing the solution, as the closest analogue to the developed composition for dust suppression on the roads of the quarry, the pre-patent was selected (Kasymkanova 2016) where the formulation developed contains a composition of wt %: tails of concentrators 35–40; cement 20–25; divinyl styrene latex 5–6; water the rest.

The basis is the task of creating a composition for fixing the dusting surfaces of the roads of the quarry, using the waste from the mining and processing industry, the coating of which has high strength and weather resistance.

To achieve this result, a composition is obtained for fixing the dusting surfaces of motor roads containing tailings of concentrating plants, water, further comprising cement and styrene–acrylic latex at the following component ratio %: tailings of concentrating factories 37.5–44; cement 22–26; styrene–acrylic latex 0.13–0.2 water the rest.

The above ratio of components was obtained experimentally and is optimal, on the one hand, to achieve the required fluidity of the solution, so that it can be easily applied to the dusting surfaces and adhesion to the dusting surface, and on the other hand, to obtain the required strength coating after setting it with the dust particles and weather-resistant.

The increase in the number of tailings of enrichment is more than 45% and reduces the fluidity of the solution and its adhesion to particles of the dusting surface with a slight increase in strength. A decrease in the amount of less than 37% increases the prime cost of the coating.

An increase in the cement of more than 26% reduces its fluidity, and a reduction of less than 22% leads to a reduction in the adhesion of the composition to the dusting surface and a decrease in the strength of the coating material. Increasing the amount of styrene–acrylic latex by more than 0.2% increases the cost of coating and slightly improves the frost resistance of the coating, and a reduction of less than 0.13% leads to a reduction in the adhesion of the formulation to the dusting surface and a decrease in the strength of the coating material (Fig. 6.1 and Table 6.2).

The following raw materials are used to prepare the composition: portland cement M 400; tails of enrichment of the Zyryanovsk ore mining and processing enterprise (further ZGOK); Styrene–acrylic dispersion (latex liquid), production of Russia.

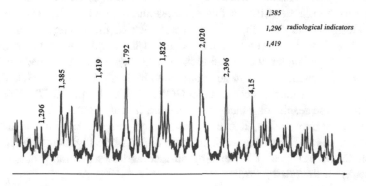

Fig. 6.1 Radiographic analysis of tailings of ore mining and processing enterprise Zyryanovsky

Table 6.2 Granulation composition of tails of ore mining and processing enterprise Zyryanovsky

Sieve sizes, mm	2.5	1.25	0.63	0.315	0.14	Less 0.14	The size module
Complete residues, %	0.1	0.2	1.0	0.30	28.0	73.5	0.4

Table 6.3 The composition of the solution

Composition	Amount mass%		
	Sample 1	Sample 2	Sample 3
Cement	22	23.8	26
Enrichment tailings	44.4	40.9	37.55
Styrene–acrylic dispersion	0.2	0.18	0.13
Water	33.4	35.1	36.3

Table 6.4 Physical and mechanical properties of the material obtained

Samples	Frost resistance, cycles	Strength, MPa, compression
1	16	22.5
2	15	24.8
3	13	27.4

Styrene and esters of acrylic acid are copolymerization products. On the basis of styrene–acrylic dispersion, various binders are developed and produced. Elasticity, vapor permeability, excellent adhesion to various substrates (adhesion), the ability to penetrate deep into the base material and thus strengthen it, resistance to weathering.

On the X-ray diffraction pattern of the tailings of ZGOK (Fig. 3.2). There are basically three components with diffraction lines 4.15; 2.396; 2.340; 2.020; 1.826; 1.792; 1.451; 1.385; 1.296 corresponding to quartz.

After dosing the components, the cement, enrichment tails, and styrene–acrylic dispersion are loaded into the concrete mixer and mixed thoroughly, with the addition of water. Examples of formulations are given in Table 6.3. The solution is delivered to the places of the dusting surfaces and applied by the spray gun.

To determine the strength and frost resistance of the obtained compositions, samples $4 \times 4 \times 4$ cm were molded. After 24 h, samples were taken from the molds and stored under wet conditions for 28 days. And then we carried out physical and mechanical tests in accordance with GOST 10180 and GOST 12730, 3, and for frost resistance—according to GOST 10060, 1 and GOST 10060 by the base method, the results of which are presented in Table 6.4.

From the above studies, it can be concluded that the use of cement in the composition makes it possible to obtain a coating having high strength and higher adhesion to the particles of the dusting surface and to solve the problem of dust suppression on the quarry roads.

References

Bek ASh, Nurpeisova MB, Turdakhunov MM (2006) Method of creating a structural-geometric model of a device array using GIS-technologies. Kazakhstan Industry 3:23–24. (in Russian)

Bekbergenov D, Jangulova G, Kassymkanova KM (2016) The possibilities of applying the technology of ore caving in caved-in deposits. Australian Mining 1:1–8. (in Russian)

Demin AM, Aleksandrov BK (1983) Deformation of ledges and sides in quarries. Moscow State University, 125p. (in Russian)

Eliseeva II, Rukavishnikov VO (1977) Grouping, correlation, pattern recognition. Statistics, 143p. (in Russian)

Kasymkanova KM, Jangulova GK, Bektur BK, Turekhanova VB (2016) Geomechanical estimation of a mountain massif in difficult mountain-geological conditions. In: Collection of works of the 2nd international scientific school of academician Trubetskoi. Problems and prospects of integrated development of subsoil. pp 77–83. (in Russian)

Kassymkanova KM (2016) Altitudinal Zonation of Exomorphogenesis in Northern Tien Shan IEJME. Math Educ 11(7):1987–2001. (in Russian)

Kasymkanova KM, Zhardaev MK et al (2010) Prepatent No. 19327 Ground reference used for observations of earth surface displacements. (in Russian)

Kasymkanova KM, Jangulova GK (2016) Investigation of the influence of structural-tectonic features and physicomechanical properties of rocks on the stability of slopes taking into account the time factor and mass explosions. Rep Natl Acad Sci Repub Kazakhstan 3:36–44. (in Russian)

Kurmankojayev AK (2003) Research of the stability of the sideboards of the Zhayremsky quarry: report on the topic No. 632/4. KazNTU; hands. KazNTU, 33p. (in Russian)

Nizametdinov FK, Portnov VS, Nizametdinov NF (2007) Modern methods of instrumental observation of the state of quarry slopes. Sanat-Printing, 77p. (in Russian)

Nizametdinov FK, Ozhigin SG, Nizametdinov RF, Obigina SB, Nizametdinov NF, Khmyrova EN (2013) Status and prospects for the development of geomechanical support for open mining. Proceedings 1:338–349. (in Russian)

Nizametdinov NF (2012) Modern methods of instrumental observation of the state of quarry slopes. The manual for the specialty Mining, Geodesy and cartography. KSTU, 65p. (in Russian)

Nurpeisova MB, Kasymkanova KM, Kyrgizbaeva GM et al (2002) Methodical recommendations for assessing the stability of the sides of the Akzhal quarry. KazNTU, 27p. (in Russian)

Ozhigin SG (2009) Control of the stability of the instrumental massifs in the quarries of Kazakhstan. Sanat printing, 44p. (in Russian)

Pevzner ME (1992) Deformation of rocks in quarries, Nedra, 235p. (in Russian)

Report on the topic, №1.758.03 (2003) Estimation of the stability of the sides of the quarries of the Rodnikov deposit. KazNTU. 105p. (in Russian)

Rudnensky AV (2003) Adhesion properties of bitumen is one of the most important indicators of their quality. Tr. GP ROSDORNII 11:123–128. (in Russian)

Turbovich IT, Gitis VT, Maslov VK (1971) Identification of images. Deterministic-statistical approach. Science, 246p. (in Russian)

Zaslavsky IY, Zorin GL, Koloskova EC, Sivolob OM (1990) Author's certificate No. 1627714. USSR Composition for the strengthening of rocks. (in Russian)

Chapter 7
Modern Methods of Researching of the Geotechnical State of the Massif for Engineering and Planning of Open Mining Works

Establishment of optimal parameters of quarry pits for the development of minerals by the open method is necessary to ensure the safety of mining operations, the safety of facilities and construction, transport and energy communications, mining equipment, and the full extraction of minerals. In this regard, during the reporting period, we developed a methodology for estimating the stress–strain state (VAT) of a rock massif in open field development. It should be noted that in the process of mining operations in the vicinity of excavations as a result of redistribution of stress, the steady state of equilibrium of rocks is disturbed. Therefore, the assessment of the stability of the sides of quarries is closely related to the study of the geotechnical state of the rock massif (Kasymkanova et al. 2015a).

Of great interest to design, practice is the problem of the steady state of equilibrium of the rock mass near open mine workings. Studying the issue of sustainability makes it possible to understand the processes occurring in the massif and correctly assess the mining situation around the workings. The problem of the stability of elastic bodies was first considered by Euler, who later developed Lagrange. The essence of the Euler–Lagrange approach is that the loss of stability is identified with the existence of new forms of equilibrium, for solving the statistical problem (Kasymkanova et al. 2015b).

Numerous full-scale observations of the deformations of the sides of quarries and studies of deformations of slopes on models from equivalent materials have established that the instrumental array undergoes complex deformation prior to collapse: horizontal stretching, vertical compression, shear. With the coefficient of stability of the slope stability, close to the limit, and stresses that accrue the creep limit of rocks, deformation develops smoothly. When modeling slopes in the process of development of deformations before the destruction of the sideboard massifs, three periods are distinguished:

© The Author(s), under exclusive licence to Springer Nature Switzerland AG 2020
K.-K. Kassymkanova et al., *Geomechanical Processes and Their Assessment in the Rock Massifs in Central Kazakhstan*, SpringerBriefs in Earth Sciences, https://doi.org/10.1007/978-3-030-33993-7_7

(1) initial, corresponding to the time of detachment of the bead, the speed of displacement based on which depends on the tempo of its design, and after completing the design for a certain period of time they have a decaying character;
(2) a period with a constant speed of deformation;
(3) a period with a progressive speed of deformation resulting in the destruction of the massif.

Accumulation of displacements of the instrumental massif for the indicated periods of relative total displacement before destruction takes place in the following ratios: (1) 45–50%; (2) 20–25%; (3) about 30%.

Simulation of slopes on equivalent materials also established zones of distribution and concentration of deformations. At the same time, the value of the strain limit values is as follows: horizontal stretching −50.10-3, vertical compression −40-50 .10-3, shifts-80-100 .10-3.

It should be noted that these strain values were observed in the slope with a safety factor of $k = 1.04$ and in the plastic deformation stage; the duration of deformation of the slope to collapse was 2.5 h. After reaching these deformation values, the slope collapsed almost instantaneously. Similar values of the deformation measures of slopes were obtained on a number of other models.

Figures 7.1 and 7.2 show the results of processing observations of the development of the deformation process of the device massif of the slope of one of the experimental models:

– isolines of critical absolute displacements (Fig. 7.1);
– isolines of critical horizontal deformations (Fig. 7.2);
– isolines of finite vertical deformations (Fig. 7.3);
– isolines of limit shifts (Fig. 7.4).

After the establishment of sufficiently reliable values of the deformations of the slopes of the models and the size of their propagation zones, a real opportunity appeared to estimate the results of calculations of the stress–strain state of slopes, the finite element method on the basis of the developed numerical procedure (Bekbergenov et al. 2016).

After the establishment of sufficiently reliable values of the deformations of the slopes of the models and the size of their propagation zones, a real opportunity appeared to estimate the results of calculations of the stress–strain state of the slopes, the finite element method based on the developed numerical procedure.

Fig. 7.1 Isolines of critical absolute displacements of the sideboard massif of the slope according to the results of laboratory modeling

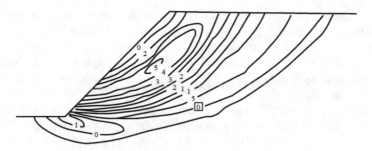

Fig. 7.2 Isolines of limiting horizontal deformations of the slope of the model

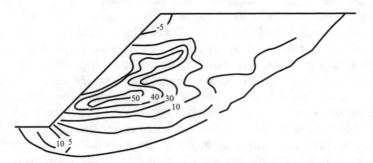

Fig. 7.3 Isolines of limiting shifts of a slope of model

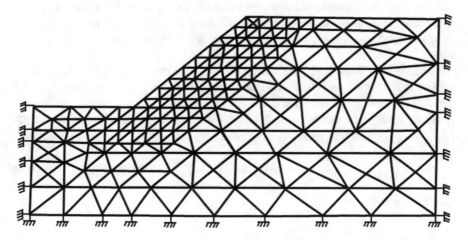

Fig. 7.4 Finite element grid

The following physical–mechanical characteristics of the mixture used in the simulation for the density $\gamma = 3.34$ g/cm^3, the angle of internal friction $\varphi = 300$, the adhesion $C = 5.85$ g/cm^2, the modulus of elasticity $E = 2 \cdot 105$ g/cm^2, the Poisson's ratio $v = 0.3$. Calculating a slope with a slope angle of 45° and a height of 64 cm with a safety factor of $k = 1.02$. The finite element scheme of the slope of the models is given in Fig. 7.4.

The basic procedure of the finite element method considers the medium as elastic and reduces ultimately to solving a system of linear algebraic equations with respect to unknown displacements: $[K]\{U\} = \{X\}$, where $\{X\}$ and $\{U\}$ are the vectors of nodal forces and displacements; $[K]$ is the stiffness matrix formed according to the known procedures of the finite element method. In our developments, the solution of nonlinear elastic–plastic problems with allowance for loosening is carried out by means of several successive linear solutions by the initial stress method. In the process of solving the problem, the load is applied by small steps in the sequence in which real loading takes place. Solutions for the next, for example, nth, load step is achieved exactly by the method of initial stresses. To the beginning of the step, the total stresses in the elements from $(n - 1)$ previous steps are known. A vector of forces and given displacements of the next stage of the load is applied to the region and elastic solutions with variable vectors are repeated in the iteration mode. To solve problems in which the main stress reverses during the loading process, it is necessary to perform stepwise loading and use a model corresponding to the principles of the theory of plastic flow. The procedure for obtaining an elastoplastic solution in the theory of plastic flow with the help of initial stresses is as follows. The load is applied by small steps in the sequence in which the real loading takes place in kind. Solutions for the next, for example, nth, load step is achieved exactly according to the above described method of initial stresses. To the beginning of the step, the total stresses in the elements from $(n - 1)$ previous steps $\{\sigma\}_{n-1}$ are known.

A vector of forces (and given displacements) of the next stage of the load is applied to the region and elastic solutions with a variable vector are repeated in the iteration mode (Kassymkhanova et al. 2016).

In the next first cycle of iterations in the elements, the deformation increment $\{\Delta\varepsilon\}_n^i$ is calculated, the corresponding elastic growth of the stresses is calculated.

$$\{\Delta\sigma^y\}_n^i = [D]\{\Delta\varepsilon\}_n^i, \tag{7.1}$$

elastic stresses

$$\{\sigma^y\}_n^i = \{\sigma\}_{n-1} + \{\Delta\sigma^y\}_n^i. \tag{7.2}$$

The "actual" voltage increase is equal to the difference between the elastic growth and the initial stresses accumulated on the previous $(n - 1)$ iterations

$$\{\Delta\sigma^{\phi}\}_n^i = \{\Delta\sigma^y\}_n^i - \{\sigma^H\}_n. \tag{7.3}$$

According to the given model of the environment, the "theoretical" increase in the stresses $\{\Delta\sigma^T\}_n^i$, the corresponding growth of the deformations $\{\Delta\varepsilon\}_n^i$ is calculated. The difference between actual and theoretical increments is considered as an increment of initial stresses

$$\{\Delta\sigma^H\}_n^i = \{\Delta\sigma^{\phi}\}_n^i - \{\Delta\sigma^T\}_n^i. \tag{7.4}$$

Initial voltage accumulates cycle after cycle within the load step

$$\{\sigma^H\}_n = \{\sigma^H\}_n + \{\Delta\sigma^H\}_n^i. \tag{7.5}$$

If the increment in the initial stresses in each of the elements does not become sufficiently small, the next $(i + 1)$th iteration begins. When the accuracy is achieved, the next $(n + 1)$th stage of the load is applied.

The above procedures for solving nonlinear problems are applicable to media both hardening and softening with associated and nonassociated state laws. All these studies further give results in the design and planning of mining operations.

By virtue of complexity of such object of planning and designing as open cast, and also because of complexity and labor intensity of the process of planning and designing, it becomes rather inconvenient to designers and technician personnel to reveal and make really optimum decisions when developing of projects and planning of mining operations in a process of deposits exploiting, using traditional methods of planning (Ozaryan et al. 2016).

In connection with integration in world information and market space began to be used in the greater degree economic parameters and criteria of estimations (CF, NPV, IRR) to planning open-cast mining, are applied methods of strategic planning, effective marketing is created aimed at maximization of the profit from mining of a deposits, there is an introduction of geostatistical methods of simulations of deposits, etc.

Increases of quality of designing and the feasibility report of accepted decisions at designing open casts can be achieved due to those advantages which are given with application of the system approach and computer facilities by the creation of effective technologies of the computer-aided designing (CAD) open mining operations. The methodology of the system approach and creation of systems of the CAD of mining operations directed on the decision the problems of the big difficult technical complexes is based on the rational decision a combination of the heuristic approach generalizing experience, intuition, and logic with numerical methods of the analysis and synthesis with attraction of methods of research of operations, acceptance of decisions of mathematical programming and simulation on the computer.

Simulating a deposit with geostatistical methods (Fig. 7.5) recently is frequently used.

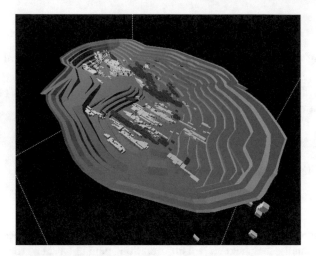

Fig. 7.5 Modeling of a contour of the open cast by triangular polyhedron and model of a part of deposits received by a geostatistical method

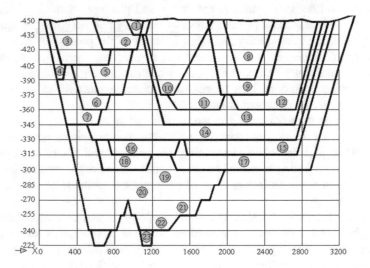

Fig. 7.6 The Circuit to a choice of final and intermediate contours career

After simulating edges of the open cast on a plane construction of volumetric triangular strips on adjacent edges and a three-dimensional model of a contour of the open cast is made (Fig. 7.6).

For a choice of rational variants final and intermediate the contours of open cast establishing with using of various principles, i.e., with use of various criteria of an estimation the subsystem provides comparison of boundary stripping ratio of a contour with average, contour, given initial plus the greatest average on the periods of mining and the current stripping ratio or rock mass. In the basis of the work of a subsystem the following calculations are made. We dissect a deposit in-depth with horizontal planes and on their lines of crossing which are checked by the designer and smooth out, in view of the given angles of walls and a design of boards (the height of edges, angles of their slopes, width of berm) in the automated mode are constructed contours of open cast in the external side before crossing with a surface and in internal—till the minimal sizes of a bottom open cast. And based on the desire of the designer on each section some variants may be constructed, including either excluding there or those ore bodies, and a bottom of the open cast may have the step form. In each variant volumes and quality of ores and rocks values of the given criterion are counted up which are compared to the boundary stripping ratio. Given step-by-step predetermined criteria on each of them, we choose a rational contour of open cast and determine the main parameters open cast (Kasymkanova et al. 2017).

The choice of a direction of development and calendar schedules of mining operations provides fixing and estimation in limits open-pit fields of a set of intermediate positions of the working zone constructed according to considered variants of the system of mining and their parameters (Fig. 7.6) (Bekmyrzaev 2013).

Intermediate positions and possible ways of development of a working zone form as multivariant graph G(EG) of a direction of development of mining operations from initial up to final the moments of development (Fig. 7.4) on which the optimum variant of development of mining operations and systems of development are found by additive criteria (set of criteria).

Let S1 (1, 2…, n)—the final set of positions of mining operations in limits of open cast field determining by the form and parameters of a working zone, constructed according to the researched system of development.

To each arch (x_i, x_j) of graph G number lij (length of an arch) is put in conformity, expressing the difference of a condition of system S in incidental tops, and each top the graph is the event. Then the problem of a determination of a rational direction of development of mining operations is reduced, firstly, to an evaluation of every ith intermediate position of mining operations; secondly, to formatting the graph of multi-variant approach of directions of development of a working zone and, thirdly, to search on graph an optimum trajectory of movement of a working zone from initial position to final. A distinctive feature of the received graph is the presence of contours (cycles).

For a determination of a rational direction of development of mining operations on graph G(EG) it is required to find the minimal (maximal) elementary way connecting initial and final nodes (Fig. 7.7).

As graph G has cycles so for each node from initial 0, which is fictitious, the optimum trajectory of movement of a working zone is calculated by the formula:

Fig. 7.7 a Columns G
(EG) multivariant approaches
of development of mining
operations, **b** Strongly
connected subgraph Gh,
c partial subgraph Gm

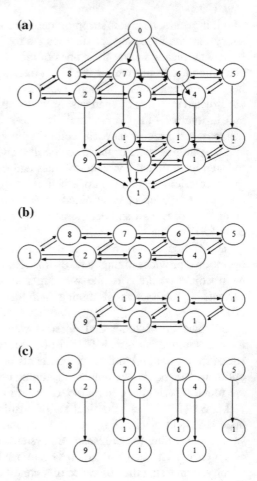

$$l_{0\pi}^* = \min(\max)\left\{ l_{0\pi}', \min_{\substack{i\in Vk \\ i\neq\pi}}(\max)l_{0i} + \left[\sum_{i1=i}^{\pi-1} li_1(i_1+1), l_{0i}' + \sum_{i1=i}^{\pi+1} l_{i1}(i_1-1)\right]\right\},$$

$$l_{0\pi}' = l_{0v}^* + l_{v\pi}; v \in V_{(k-1)}, \pi \in V_k,$$
$$\text{where } (v,\pi) \in U_{(k-1),k}, k = 1,2\ldots,(r-1),$$
$$l_{0\pi}' = l_{0\pi}, \pi \in V_1, k = 1$$

Optimization of calendar schedules is realized by means of
economic-mathematical models, which are based on methods of the decision of
integer problems of linear programming with Boolean variables and are formulated
as follows.

To minimize (to maximize) criterion function

$$F(x_j) \rightarrow \text{opt} \tag{7.6}$$

Under conditions imposed on ore output and blocks:

$$Q_p(1 - \mu_p) \leq \sum_{j \in U} Q_{p,j} x_j + \sum_{j \in U_{pH}} Q_p \leq Q_p(1 + \mu_p), p = \bar{1}p \tag{7.7}$$

$$\alpha_{p,q}^{\min} \leq \frac{\sum_{j \in U} Q_{p,j} \alpha_{pq,j} x_j + \sum_{j \in U_{pH}} Q_{p,j} \alpha_{pq,j}}{\sum_{j \in U} Q_{p,j} x_j + \sum_{j \in U_{pH}} Q_{p,j}} \leq \alpha_{p,q}^{\max}, pq \in P_q \tag{7.8}$$

$$k_b(1 - \mu_k) \leq \frac{\sum_{j \in U} v_j x_j + \sum_{j \in U_{pH}} v_j}{\sum_{j \in U} \sum_{p=1}^{P} Q_{p,j} x_j + \sum_{j \in U_{pH}} \sum_{p=1}^{P} Q_{p,j}} \leq k_b(1 + \mu_k), \tag{7.9}$$

$$x_0 = 1, i \in U_{pH}; \ |U_j| x_j \leq \sum_{i \in U_j} x_i; j \in U_a; \tag{7.10}$$

$$x_{j-1,k} + x_{j+1,k} + x_{j,k-1} + x_{j,k+1} - 3x_{j,k} \leq 1$$

$$4 x_{j,k} - x_{j-1,k} - x_{j+1,k} - x_{j,k-1} - x_{j,k+1} \leq 0 \tag{7.11}$$

where U_c-set even columns of models on axis Y and $U_c \subset U$; v_i, $Q_{i,j}$ accordingly volumes of overburden and pth kind of ore in ith element; $\alpha_{pq,j}$—qth quality index pth kind of ore in jth element; R the max possible volume of rock mass for planned year; the U—selected set of elements for planned year; Up.—a subset of the elements included to a contour when choice of a rational direction of development of mining operations; $Q_p^{\min}$, the $Q_p^{\max}$—bottom and top changes of volume pth kind of ore; $Q_{pq}^{\min}$, $Q_{pq}^{\max}$—requirements to qth quality index in pth grade of ore; μ_p, μ_k—allowable changes of volume pth kind of ore and stripping ratio; U_a-set of technologically depending elements of model $U_a \subset U$; U_j—the set of elements of model, technologically directly depend its jth an element; U_b—set of elements even columns models $U_b \subset U$; X_j—variable which is equal 1, if jth element is included in the plan tth planned year, and to zero otherwise; k_b—given stripping ratio.

The subsystem copes by the change of criterion function and restrictions. The used method is realized by set new optimum models. It is also possible to note, that the complex of programs is created on the automated formed blocks, with the help of the task of sectors and strips and the subsequent completion of the calendar schedule by the expert in an interactive graphic mode.

Optimization of freight traffic and schemes of the opening up of open cast consists of searching such transport system of the open cast from all possible variant, which at a minimum of expenses provided planned volume of rock mass transportation.

In formalized form opening workings and transport networks may be presented in a form of an oriented graph G(NL), where *L*—set of arcs and *N*—set of nodes. Nodes of graph are images of real points of loading, discharging and crossing of transport lines, which are boundaries of two or more adjacent parts of network [83-86]. Numbers r_{ij}, which are transport capacity of arcs, and function of cost **F**, which characterizes costs of driving and maintenance of a district *ij*th and also costs of rock mass transportation, made correspond to every arc *ij*th of the graph. Economic-mathematical model provides for joint optimization of open cast's opening and open cast's freight traffics, systems of limitation of goal function, ascertaining by analogy with equations of the first and the second Kirchhoffs laws for electric network, balance condition, when volume of production in points of loading during t time must be not more than transport capacity of arcs and be equal to intake capacity of points of rock mass receipt. Offered model ensured correct solving of given problems in conditions of the concrete open cast when designing and planning of mining operations [87]. The algorithm of a solution of a problem represents the construction of an allowable initial flow and consequent iterative process of sequential redistribution freight float, reducing the significance of criterion function. The redistribution of a stream is carried out in elementary cycle constructed on arcs of a tree by adding of one arc—of connection, and is finished if the survey of all cycle has not revealed a possibility of a diminution of criterion function in one from a cycle.

Determination of the optimum parameters of the quarry pits for the mining of minerals by the open method is necessary to ensure the safety of mining operations, the preservation of facilities and structures, transport, and energy communications, mining equipment, the full extraction of minerals, and finally, the high economic performance of mining enterprises.

The complex of the developed programs allows making input, survey, and adjustment of a database the graph of G(NL). On the graph it is possible to build the shortest paths, trees and to decide problems of optimization of freight traffic and schemes of the opening—up of open cast. Models for a determination of parameters of loading transport complexes of open pit are based on logic-statistical imitating modeling of mining-transport processes. Models of construction of opening inclined mining working and transport communications are based on heuristic dialogue algorithms.

Studies previously reported in various scientific conferences, journals. This article is proposed with additions and new research.

References

Bekbergenov D, Jangulova G, Kassymkanova KM (2016) The possibilities of applying the technology of ore caving in caved-in deposits. Aust Min 1:1–8. (in Russian)

Bekmyrzaev BZ (2013) Principles of systems approach in Kazakhstan geoinformation mapping. In: Seventh central Asia conference GISCA13. Kazakhstan, pp 102–106. (in Russian)

Kasymkanova KM, Jangulova GK, Bektur BK (2015a) The mineral and raw material complex of Kazakhstan is the basis of social and economic development. Vestnik, geographic series, KazNU 1:165–171. (in Russian)

Kasymkanova KM, Jangulova GK, Baidauletova GK, Zhalgasbekov YZ (2015b) Waste of mining production. Vestnik, geographic series, KazNU 1:173–186. (in Russian)

Kasymkanova KM, Jangulova GK, Bektur BK, Turekhanova VB (2016) Geomechanical estimation of a mountain massif in difficult mountain-geological conditions. In: Collection of works of the 2nd international scientific school of academician Trubetskoi. Problems and prospects of integrated development of subsoil, pp 77–83. (In Russian)

Kasymkanova KM, Jangulova GK, Turekhanova VB (2017) Investigation of the geomechanical state of the mountain massif in difficult mining conditions. In: Materials of XXYII international scientific school named after. Academician SA Khristianovich, Crimea, pp 117–120. (in Russian)

Ozaryan YA, Savostin IV, Jangulova GK (2016) Application of geoinformation technologies for environmental and predictive assessment of the environmental quality of the territory of mining development, vol 21(Special Issue). Gornaya Kniga Publishing House, pp 493–497. (in Russian)

Conclusion

The monograph is devoted to methodology, to a new approach to developing and improving the stability of the quarry sides, new ways to increase the stability of the slope of the ledges by strengthening and strengthening them, is the result of three years of research.

In the course of the work, an analysis of the world experience in the study of the geomechanical state of the mountain massif in complex mining-geological and mining conditions during the mining of mineral deposits by the open method is carried out. Based on the analysis, it can be concluded that mineral deposits developed by the open method are characterized by a great variety of mining, geological, mining, geomechanical and technological conditions.

The influence of structural-tectonic features and physicomechanical properties of rocks on the stability of slopes with the factor of time and mass explosions was investigated. From the variety of factors affecting the stability of slopes with enclosing rock and semi-rock rocks, three main factors have been identified that require compulsory registration in the study of geomechanical processes.

Based on the study of the fracture of the tectonics of the deposit, developed by the open method, the types of work were determined: field work, drawing up and processing of point and other diagrams based on field measurements to identify crack systems; study of physical and mechanical properties of rocks; laser scanning; geophysical research. It is recommended to conduct local researches of a pit slope placed in the limit position, in the area of which drilling and blasting operations are planned, and to find out the depth of disturbance (cracks, shears, discontinuities, etc.) using the method of thermometry.

A number of repeated experimental studies using the method of thermometry were carried out on various rock samples both integral and with fine fracturing. Dependences of the output voltage of the photodetector on the temperature of the sample surface—emitters for different rocks are obtained.

To prevent deformation of rocks, artificial reinforcement is proposed, which makes it possible to provide the necessary stability of the slopes of the ledges of the non-working sides of quarries and in some cases to prevent possible collapse of rocks in weakened areas, in others—to significantly reduce the amount of stripping work.

A distinctive feature of the work is that new methods for developing the composition of cementitious solutions for hardening and strengthening pit slopes have been explored and proposed in detail.

Printed in the United States
By Bookmasters